Lecture Notes in Computer Science 13659

Katja Gilly · Nigel Thomas (Eds.)

Computer Performance Engineering

18th European Workshop, EPEW 2022
Santa Pola, Spain, September 21–23, 2022
Proceedings

Editors
Katja Gilly (iD)
Miguel Hernandez University
Elche, Spain

Nigel Thomas (iD)
Newcastle University
Newcastle upon Tyne, UK

ISSN 0302-9743   ISSN 1611-3349 (electronic)
Lecture Notes in Computer Science
ISBN 978-3-031-25048-4   ISBN 978-3-031-25049-1 (eBook)
https://doi.org/10.1007/978-3-031-25049-1

This Springer imprint is published by the registered company Springer Nature Switzerland AG
The registered company address is: Gewerbestrasse 11, 6330 Cham, Switzerland

Preface

This volume of LNCS contains papers presented at the 18th European Performance Engineering Workshop held in Santa Pola, Alicante, Spain in September 2022.

The accepted papers reflect the diversity of modern performance engineering. The sessions covered a wide range of topics including robustness analysis, machine learning, edge and cloud computing, as well as more traditional topics related to stochastic modelling, techniques and tools. This was the first time that EPEW was held in person since the Covid-19 pandemic. While the overall number of papers and presentations was lower than some previous events, it was nevertheless an important step in renormalising workshop participation and sharing high-quality work in a positive and productive environment. A total of 14 full papers were accepted in the Workshop after going through a single blind revision process that guaranteed a minimum of 3 reviews for each paper.

We were delighted to have keynote presentations from Dr. Xavier Costa Pérez from NEC Laboratories Europe and Prof Aad van Moorsel from the University of Birmingham. These talks reflected the state of performance engineering today. Xavier Costa Pérez's talk concerned AI for Future 6G Systems, covering AI-driven automation for industry verticals, O-RAN network disaggregation, smart surfaces and wireless sensing. Aad van Moorsel's talk on Probabilistic Models for Blockchain presented a number of results giving insight into the behaviour and challenges of blockchain systems.

As workshop co-chairs we would like to thank everyone involved in making EPEW 2022 a success: Springer for their continued support of the workshop series, the programme committee and reviewers, and of course the authors of the papers submitted, without whom there could not be a workshop. We would especially like to extend our thanks to the staff of the Museo del Mar at the Castillo Fortaleza, Santa Pola, for hosting the workshop and providing an excellent environment and support. We trust that you, the reader, find the papers in this volume interesting, useful and inspiring, and we hope to see you at future European Performance Engineering Workshops.

November 2022

Katja Gilly
Nigel Thomas

Organization

Chair

Katja Gilly   Miguel Hernández University (UMH), Spain

Organizing Committee

Salvador Alcaraz   Miguel Hernàndez University (UMH), Spain
Cristina Bernad   Miguel Hernàndez University (UMH), Spain
Pedro Pablo Garrido   Miguel Hernàndez University (UMH), Spain
Eduardo López   Miguel Hernàndez University (UMH), Spain
Pedro J. Roig   Miguel Hernàndez University (UMH), Spain

EPEW 2022 Program Committee

Chair

Nigel Thomas   Newcastle University, UK

Members

Noura Aknin   Abdelmalek Essaadi University, Morocco
Salvador Alcaraz   Miguel Hernandez University, Spain
Elvio Gilberto Amparore   Università degli studi di Torino, Italy
Paolo Ballarini   CentraleSupelec, France
Enrico Barbierato   Università Cattolica del Sacro Cuore, Italy
Marco Beccuti   Università degli studi di Torino, Italy
Marco Bernardo   University of Urbino, Italy
Laura Carnevali   University of Florence, Italy
Ioannis Dimitriou   University of Patras, Greece
Dieter Fiems   Ghent University, Belgium
Sonja Filiposka   University of Ss. Cyril and Methodius, North Macedonia
Matthew Forshaw   Newcastle University, UK
Jean-Michel Fourneau   University of Versailles, France
Pedro Pablo Garrido   Miguel Hernandez University, Spain
Marco Gribaudo   Politecnico di Milano, Italy
Boudewijn Haverkort   Tilburg University, Netherlands
András Horváth   University of Turin, Italy

Esa Hyytia	University of Iceland, Iceland
Mauro Iacono	Università della Campania Luigi Vanvitelli, Italia
Alain Jean-Marie	Inria, France
Carlos Juiz	University of the Balearic Islands, Spain
William Knottenbelt	Imperial College London, UK
Lasse Leskelä	Aalto University, Finland
Andrea Marin	University of Venice, Italy
Anastas Mishev	University of Ss. Cyril and Methodius, North Macedonia
Nihal Pekergin	Université Paris-Est Créteil, France
Tuan Phung-Duc	University of Tsukuba, Japan
Agapios Platis	University of the Aegean, Greece
Anne Remke	WWU Münster, Germany
Markus Siegle	UniBw Munich, Germany
Miklos Telek	Budapest University of Technology and Economics, Hungary
Joris Walraevens	Ghent University, Belgium
Katinka Wolter	Frei Universität Berlin, Germany

Contents

Robustness Analysis

Measuring Streaming System Robustness Using Non-parametric Goodness-of-Fit Tests

Stuart Jamieson[✉] and Matthew Forshaw

Newcastle University, Newcastle upon Tyne NE1 7RU, UK
{S.Jamieson3,matthew.forshaw}@ncl.ac.uk

Abstract. Due to unpredictable disturbances in the operating environment, stream processing systems may experience performance degradation and even catastrophic failure. Streaming systems must be robust in the face of such uncertainty in order to be deemed fit for purpose. Measuring and quantifying a system's level of robustness is a non-trivial task. We present, compare and contrast a range of non-parametric goodness-of-fit tests which can act as quantifiers of a system's level of robustness. We show that different tests produce differing relative measures of system robustness, affected by not only the test statistics inherent characteristics, but also by the particular latency percentile under scrutiny.

Keywords: Streaming system · Robustness · Testing · Non-parametric

1 Introduction

Streaming "big data" forms a vital part of many companies' business infrastructure. Many systems and applications in use today involve a large volume of continuous data streams along with large-scale, diverse, and high-resolution data sets. Unpredictable circumstances, such as a higher than expected system load or variability, can cause parallel and distributed streaming systems to experience performance degradation [10,12,14–16,21]. It is possible for a system to perform to acceptable levels under normal operating circumstances and yet exhibit catastrophic failure when subjected to slight disturbances [4].

A minor degradation in application performance can have a high penalty on system operators. Google reported a 20% revenue loss when an experiment caused an additional delay of 500 ms in response time [19]. Amazon reported a 1% sales decrease for an additional delay of 100 milliseconds in returning search results [18]. On June 29th 2010, Amazon.com suffered a 3-h period of intermittent performance problems, including high page latency and incomplete product search results. This led to a loss of $1.75m in revenue per hour [24].

Due to these high penalties for system performance degradation, streaming system operators may seek to accept lower average performance in return for an increase in robustness to a wide variety of operating conditions [5]. This raises the question of how robustness is to be quantified and measured. We evaluate several robustness metrics based on non-parametric goodness-of-fit tests. Our aim is to identify emerging best practice which could then inform future performance analysis research considering robustness quantification in distributed systems.

2   Background and Motivation

A robust system is one that can maintain performance under a wide variety of operating conditions, or in the face of various uncertainties in the operating environment. Robustness is the persistence of certain specified features, despite the presence of perturbations in the system's environment [5].

Our work is most closely related to [5], which presents a robustness metric based upon the *Kolmogorov-Smirnov* (KS) statistic. The suitability of KS arises from the fact that it can be used in any analytical model for which the cumulative distribution functions (CDFs) of the disturbance and performance can be estimated. It can also be applied to the empirical CDFs (ECDFs) of actual system data. Given a set of performance observations, if the system is robust, the ECDF of the performance metric under "normal" conditions $F(x)$, should be very similar to the ECDF of the performance metric with perturbations applied $F^*(x)$. The closer the functions agree, the more robust the system.

To analyse the persistence of relevant specified features, e.g., system latency, data must be collected across a range of environments, i.e., situations whereby certain disturbances are applied and those without. The results can then be compared and conclusions drawn as to the magnitude of effect each disturbance has on the chosen performance metric. When measuring data from complex systems, one often has no a priori knowledge regarding the underlying distribution function from which the data originate. To avoid making assumptions as to the underlying distribution, non-parametric tests, as opposed to parametric tests, are most appropriate when comparing two data samples for goodness of fit.

Although appropriate for use in our case, the KS statistic also displays certain characteristics which suggest it may not be the optimal approach in certain circumstances. For example, while the KS statistic is robust to outliers, it is known that the values of the KS statistic itself are not equally sensitive to movements along its own probability distribution. The KS statistic is most sensitive to situations where ECDFs differ in a global fashion near the centre of the distribution, but is relatively insensitive to differences at the extremes [2]. This is due to values converging to zero and one, at the extremes of the ECDFs.

This poses two problems; firstly, real-world computing systems are known to experience "heavy-tailed" distributions of metric values when experiencing performance disturbances or degradation [11]. Secondly, observations near the right tail of the ECDF, by definition, represent more extreme values of the underlying metric than those observed nearer the median. More extreme values result in more extreme impacts upon the system. The insensitivity of the KS statistic in the tail regions goes against this logic by implicitly giving less weight to these extreme observations than they deserve.

There exist numerous other non-parametric goodness-of-fit tests from the same family of tests as the KS test, each created to better cope with different sample sizes or circumstances. We propose incorporating a number of these statistical tests, along with the KS statistic on the basis that the resulting values either include additional information not provided by the KS statistic alone, or overcome a particular weakness or assumption within the other tests.

Our analysis shows that different tests will produce differing results to provide a more complete picture of the 1) level and 2) characteristics of a streaming system's robustness. A streaming system's robustness can be characterised by its reaction to various levels of disturbance, rather than to a single level. We relate a robust system to one that shows a size and rate of degradation that is in line with the size and rate of change of the applied disturbances. If this is the case then the test statistic value will follow a predictable, near-linear path in the face of changes in the size of disturbance. Vice versa we relate a sensitive, non-robust system to one that shows large, possibly non-linear changes in the chosen test statistic in relation to the applied disturbance.

3   Summary of Test Statistics

Within this section, we formalise notation for each test statistic under evaluation.

3.1   Kolmogorov-Smirnov

The Kolmogorov-Smirnov (KS) test [22] belongs to the supremum class of empirical distribution function (EDF) statistics and is based on the cumulative probability distribution of data. The *one-sample KS test* is used to compare a sample distribution with a reference probability distribution to decide if the sample comes from a population with that specific distribution.

The *two-sample KS test* is used to compare two sample distributions to decide whether they were drawn from the same (but unknown) population. It calculates the distance between the EDFs of the two samples, at each unit of the scale, e.g., at each time point for latency distributions, where the value of interest is the maximum vertical distance identified. For the two-sample KS test, the null hypothesis is that both sample distributions come from the same underlying distribution. If the KS statistic $\delta_{nn'}$, shown in Eq. 1, is larger than the critical value, the null hypothesis is rejected.

$$\delta_{nn'} = \sqrt{\frac{nn'}{n+n'}} \sup_x |F_n(x) - F_{n'}(x)| \tag{1}$$

where $F_n(x)$ is the ECDF of data size n and $F_{n'}(x)$ is the ECDF of data size n'.

Due to the insensitivity of the KS statistic to differences in the tails of a distribution, a weighting function ψ is applied to adjust it, creating the weighted-KS statistic (WKS) δ_w, in order to assign higher weights to larger values in the right tail of the ECDF [5]. This is shown in Eq. 2.

$$\delta_w = \delta_{nn'}\psi(x) \tag{2}$$

The weighting function applied needs to account for the underestimation of the KS statistic in the right tail, but not overly so. Previous experimentation carried out by [5] suggests choosing the function shown in Eq. 3 to apply in the computation of δ_w. This function is based on the fact that the quantity $F(x)(1 - F(x))$ is at a maximum where $F(x) = 0.5$, and the weight is only applied when δ occurs to the right of the median, otherwise $\psi(x) = 1$.

$$\psi(x) = -\ln(F(x)(1 - F(x))) \qquad (3)$$

3.2   Cramér-von Mises

The *Cramér-von Mises* (CVM) statistic [20] belongs to the quadratic class of EDF statistics and is defined as the integrated squared difference between the EDFs of the two samples being compared. If the value of T, calculated as shown in Eq. 4, is larger than the tabulated values, the hypothesis that the two samples come from the same distribution can be rejected.

$$T = \left[\frac{NM}{(N + M)}\right] \int_{-\infty}^{\infty} [F_N(x) - G_M(x)]^2 \, dH_{N+M}(x) \qquad (4)$$

where $F_N(x)$ and $G_M(x)$ are the ECDFs for the first and second sample, $H_{N+M}(x)$ is the ECDF for the two samples, and N and M are the sample sizes.

While the KS statistic can be insensitive to distributions with similar means that display multiple "cross-overs" in their ECDFs, the CVM statistic retains its sensitivity in this situation as it measures the sum of squared differences, not just the differences themselves. However, the CVM statistic is insensitive to differences in the tails of the distributions, in similar fashion to the KS statistic.

3.3   Anderson-Darling

The *Anderson-Darling* (AD) test [2] belongs to the quadratic class of EDF statistics and was developed originally for detecting sample distributions' departure from normality. The two-sample AD test states the null hypothesis that the two samples come from the same continuous distribution and is rejected if the AD statistic is larger than the corresponding critical value. The two-way AD statistic is calculated as per Eq. 5.

$$AD = \frac{1}{mn} \sum_{i=1}^{n+m} \left(N_i Z_{(n+m-ni)}\right)^2 \frac{1}{iZ_{(n+m-1)}} \qquad (5)$$

where Z_{n+m} is the ECDF for the two combined and ordered samples X_n and X_m, of size n m respectively, and N_i represents the number of observations in X_n that are equal to or smaller than the ith observation in Z_{n+m}.

Analysis carried out by [6] compared the AD test with the KS test and found the AD test more powerful when comparing two distributions that vary in shift only, in scale only, in symmetry only, or that have the same mean and standard deviation but differ in the tail ends only.

3.4 Epps-Singleton

The *Epps-Singleton* (ES) test [7] compares the empirical characteristic functions (ECFs) of two samples, rather than the observed distributions. The *p*-value of the ES test gives the probability of falsely rejecting the null hypothesis that both samples have been drawn from the same population. The ECF is defined as the Fourier transform of the distribution function, as shown in Eq. 6. The ECF is known to completely characterise any distribution and can be used to derive its moments, so contains more information than just a single measure such as the mean or variance.

$$\phi_{n_k}(t) = \int_{-\infty}^{\infty} e^{itx} dF_{n_k}(x) = n_k^{-1} \sum_{m=1}^{n_k} e^{itX_{km}} \tag{6}$$

where ϕ is ECF, t is a real number, i is $\sqrt{-1}$, n is sample size and X_{km} is the m-th observation in sample k.

[7] compared the ES test with the AD, CVM and KS tests resulting in the following conclusions [9]; a) Apply the ES test if using discrete data, b) the KS test usually has a lower power than the ES test if using continuous data, and c) sometimes the AD and CVM tests can have a higher power than the ES test.

4 Methods

This experiment aims to measure changes in streaming system performance, in the face of disturbances, i.e., changes in the characteristics of the incoming workloads. A range of workloads were generated and run using Flink [3]. Latency metrics were selected as our proxy with which to represent performance, or any degradation thereof. Having recorded the latency observations across the different workload simulation runs, the results were analysed and transformed to allow an indication of how robust the system was to the changing workloads.

4.1 System

The evaluation was carried out using Apache Flink 1.4.1 [3] and Nexmark [1,23] benchmark suite. Experiments ran on an X570 AORUS ULTRA with AMD Ryzen 9 3900X 12-Core Processor and 64GB RAM, running Microsoft Windows 10 Pro. Data collection was integrated with SLF4J in a Docker container.

The workloads were run through a three-step *Word Count* topology with three unique operators. In our topology, the *Source Operators* emit sentences of a particular word-length by randomly selecting from a provided set of English words. The sentences are sent to the second stage of the topology, where the *Splitter Operators* split them into individual words. These words are, in turn, sent forward to the third stage of the topology, the *Count Operators*. At this stage, a count is kept of the number of times each word was encountered.

4.2   Workloads

We design our workloads by specifying the workload generation function and the input variables which are to be varied to simulate the necessary disturbances. We create a range of simulations by varying the inputs to the chosen workload function, using a *one-factor-at-a-time* (OFAT) experiment design approach [8].

The workload generation function is a sinusoidal wave (Eq. 7) where a is amplitude, ω is angular frequency, ϕ is phase shift and δ is vertical shift.

$$Y(t) = a\sin(\omega(t + \phi)) + \delta \tag{7}$$

The input variables to be varied are: 1) The number of source operators generating the incoming workload, 2) the amplitude of the sine wave workload generation function vs its vertical distance, 3) The angular frequency of the sine wave workload generation function, and 4) The sentence size on the Nexmark Word Count workload logic. The workload variations, driven by the OFAT approach, are shown in Table 1. 11, equally spaced levels were chosen for each input value to be varied, resulting in 44 simulation runs carried out in total.

Table 1. OFAT workload design

Input variable	Input chosen to vary			
	Source operators	Amplitude	Sentence size	Angular frequency
Source operators	$\{5, 6...15\}$	10	10	10
Splitter operators	10	10	10	10
Count operators	10	10	10	10
Amplitude	10,000	$\{1000k | k \in \{5, 6...15\}\}$	10,000	10,000
Angular frequency	0.006978	0.006978	0.006978	$\{\frac{1}{120k}2\pi | k \in \{5, 6...15\}\}$
Sentence size	100	100	$\{10k | k \in \{5, 6...15\}\}$	100
Vertical shift	20,000	20,000	20,000	20,000

5   Results

Source Operator Variability. Figure 1 shows the ECDFs of the max (1a), 99th (1b), 95th (1c) and 50th (1d) percentile latencies for each Flink simulation, for which the number of workload source operators were varied. There are noticeable deviations in the left tail for the three highest percentile latencies while, although all latency percentile ECDFs have extreme right tails, significant deviations between the ECDFs within those extreme right tails only occur for the max and 99th percentile latencies.

Furthermore, as the latency percentiles increase, the extreme right tail values account for larger proportions of the observed values, i.e., the maximum values for each percentile are not significantly different, but we see a significantly different number of observations at or near those maximum values. For example, using a 250 ms latency value as the threshold, we observe proportions of total observations above that value of; circa 1% for 50th percentile latencies, circa

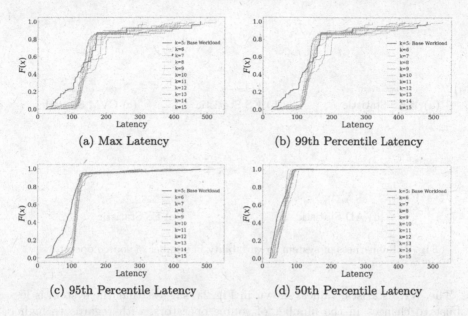

(a) Max Latency

(b) 99th Percentile Latency

(c) 95th Percentile Latency

(d) 50th Percentile Latency

Fig. 1. ECDFs of percentile latency values with variable source operators

Table 2. Test stat correlation across latency %tiles w. variable source operators

Statistic	Max					p99					p95					p50				
	WKS	KS	AD	ES	CVM	WKS	KS	AD	ES	CVM	WKS	KS	AD	ES	CVM	WKS	KS	AD	ES	CVM
WKS	1.0	0.9	0.88	0.35	0.9	1.0	0.91	0.92	0.46	0.94	1.0	1.0	0.95	0.91	0.95	1.0	1.0	0.94	0.82	0.95
KS		1.0	0.94	0.49	0.91		1.0	0.95	0.52	0.92		1.0	0.95	0.91	0.95		1.0	0.94	0.82	0.95
AD			1.0	0.55	0.98			1.0	0.59	0.95			1.0	0.98	1.0			1.0	0.85	1.0
ES				1.0	0.53				1.0	0.58				1.0	0.98				1.0	0.84
CVM					1.0					1.0					1.0					1.0

3–4% for 95th percentile latencies, circa 10–12% for 99th percentile latencies, and circa 15% for max percentile latencies.

Figure 1 also shows that the higher percentile latencies (max and 99th) display significantly more instances of ECDF cross-overs, than do the ECDFs for the lower percentile latencies (95th and 50th). As the latency percentile in question increases, it becomes more important to capture any differences in the right tail of the distribution as those differences become larger, along with retaining sensitivity to any instances of ECDF cross-over.

Table 2 shows that the 50th and 95th percentile latency results display higher overall levels of inter-statistic correlation than those observed for the max and 99th percentile latencies. This is in keeping with expectations; the more sensitive a statistic is to non-locational, non-global differences in the distributions, (e.g., the more sensitive it is to differences at the extremes or to multiple ECDF cross-overs), the more it should begin to differ, in this case, from other statistics which are less sensitive to these characteristics.

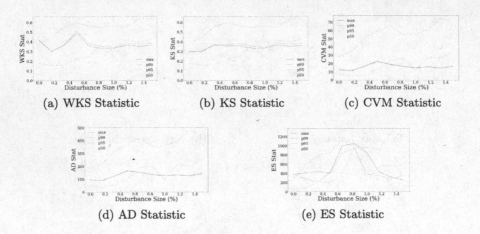

(a) WKS Statistic   (b) KS Statistic   (c) CVM Statistic

(d) AD Statistic   (e) ES Statistic

Fig. 2. Robustness of system to variability in number of source operators

The WKS statistic values, shown in Fig. 2a, suggest that the system is less robust to changes in the number of source operators, with regards to higher latency percentiles, than do the KS statistic values shown in Fig. 2b. The WKS statistic values however suggest the system is as equally robust to changes in the number of source operators, with regards to the lower percentile latencies as do the KS statistic values. This appears to show the WKS statistic working as intended to afford more attention to values occurring in the right tails. A system experiencing high values of high percentile latency observations represents a system already under potential stress, or near the bounds of its "stable operating environment" and could be considered more likely to lack robustness.

The CVM and AD statistics (Figs. 2c and 2d), represent the system as being significantly more robust in the max and 99th percentile latencies than for the 95th percentile latency. The WKS statistic (Fig. 2a) shows a markedly smaller difference in robustness between the 95th and the higher percentile latencies. This is expected behaviour as the CVM statistic is known to be insensitive to deviations at the tails, which is the case here for the higher percentile latencies. This would result in a CVM statistic value which deems the robustness level for the various percentile latencies to be more similar to that shown by the KS statistic (Fig. 2b), as it does not differentiate as markedly between ECDFs with and without larger deviations in the tails, as does the WKS statistic.

The behaviour of the AD statistic is less expected. The AD statistic is said to retain sensitivity to deviations in distributions, even when those deviations occur in the extreme tails. For this reason, one might expect the AD statistic to show a difference in system robustness levels that is more similar to that shown by the WKS statistic, i.e. it would capture the lack of robustness implied by any deviations occurring at the extremes of the ECDFs, and deem the system to be less robust in terms of the max and 99th percentile latencies, therefore shifting their statistic values higher and closer to those of the 95th percentile latency. Considering this, whereby the AD statistic is more similar to the KS than the

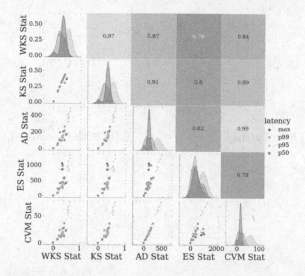

Fig. 3. TS pair plot across all latency percentiles for variable source operators

WKS statistic, we may deem the AD statistic to be potentially under-estimating the sensitivity of the system's high percentile latency distributions to changes in the number of source operators, (i.e. over-estimating the levels of robustness).

The ES statistic suggests an even smaller difference between the system robustness across different percentile latencies, with Fig. 2e displaying a number of observed instances whereby the max and 99th percentile latency distributions are less robust than the 95th percentile latency distributions to certain disturbances in the number of system source operators. This is in contrast to the KS, WKS, CVM and AD statistics which represent the 95th percentile latency distributions to be less robust than the higher percentile latencies across the entire range of disturbances to the number of source operators within the system. The ES statistic also appears to suggest a substantially lower level of system robustness overall for the higher percentile latencies, when compared to the CVM and AD statistics. Figure 2e displays a non-linear, somewhat erratic relationship for the max, 99th and 50th percentile latencies.

Figure 2 shows much more agreement between the various statistics when considering the robustness of the system for the lower percentile latencies. This implies that the statistics may capture differences in the distributions of the higher percentile latencies and the more extreme values in ways that differ more than when capturing the effects of less extreme values and differences in the lower latency percentile distributions. At a high level, the majority of the test statistics in this case, (save for the ES statistic) imply the system to be relatively more sensitive to initial, smaller disturbances (higher latencies more so than lower percentile latencies), becoming less sensitive to larger disturbances, e.g. the max and 99th percentile latency test statistic values reach their maximum level of deviation after disturbance sizes of between 75–100%, levelling out thereafter.

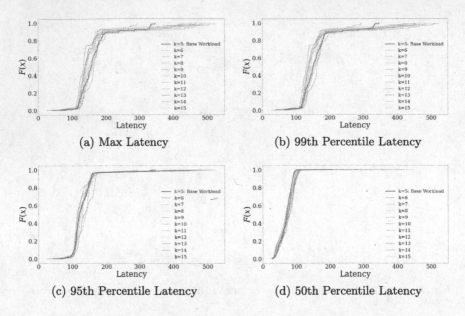

Fig. 4. ECDFs of percentile latency values with variable frequency

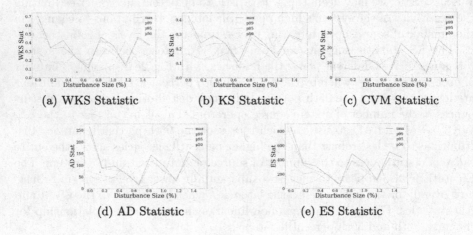

Fig. 5. Robustness of system to variability in frequency

This higher percentile lack of robustness to initial disturbances (i.e. disturbances across the lower half of the total range of values tested) may be due to the fact that those latency percentiles already represent extreme values observed, most often when a system is under strain. If a system is already under strain and at the extremes of its "normal" stable operating region, then a small disturbance can cause a significant reaction and degradation. The system may then become relatively less sensitive to larger disturbances as things can "only get so bad" before the system fails. In turn the 95th percentile latency appears the least robust metric according to the majority of test statistics. Further investigation is required, however It may be due to the values in the 95th percentile being located relatively deep into the right tail, at a point where the underlying characteristics of the distribution begin to change, e.g. Extreme Value Theory (EVT) begins to govern rather than the Central Limit Theorem (CLT).

We see from Fig. 3 that when considering the 50th and 95th percentile latencies, the ES statistic generally displays a positive, linear relationship with the other statistics. Table 2 shows that for the max and 99th percentile latencies, however, the ES statistic values display much lower levels of correlation between it and the other statistics' values. This may suggest that the ES statistic is more sensitive to differences between extreme right tail values, as the ES statistic continues to vary, for the higher percentile latencies, beyond a point at which the other statistics begin to remain almost static.

Frequency Variability. Similarly to when the number of source operators is varied, when the angular frequency input to the Sinewave workload generation function is varied, the resulting ECDFs, shown in Fig. 4, display significant deviations in the right tails, when considering the higher percentile latencies (max and 99th). In contrast, however, there appear to be no significant deviations located in the left tail for any of the percentile latencies.

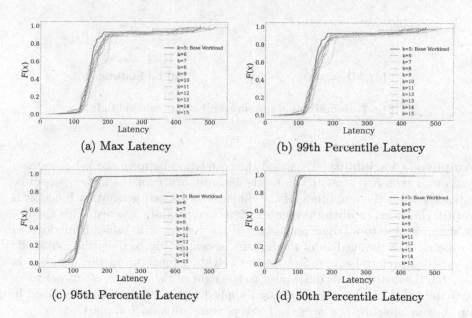

(a) Max Latency

(b) 99th Percentile Latency

(c) 95th Percentile Latency

(d) 50th Percentile Latency

Fig. 6. ECDFs of percentile latency values with variable amplitude

The WKS statistic values, shown in Fig. 5a, suggest that the system is less robust to changes in angular frequency across all latency percentiles than do the

KS statistic values. Rather than the WKS statistic weighting function affecting the higher percentile latencies' values more noticeably than those of the lower percentile latencies, we see more of a linear, upward shift for all values, as shown in Fig. 5a versus Fig. 5b. This linear shift maintains the nature of the relationship between the WKS statistic values and the KS statistic values across the range of angular frequency disturbances tested. This relationship between the test statistic values for the differing latency percentiles across various levels of disturbance in angular frequency appears to hold true across the remaining test statistics (Figs. 5c 5d 5e). All statistics suggest that the system is least robust to small changes in the angular frequency when considering the max and 99th percentile latencies. As the size of the disturbance increases, the levels of robustness converge across the different percentile latency values.

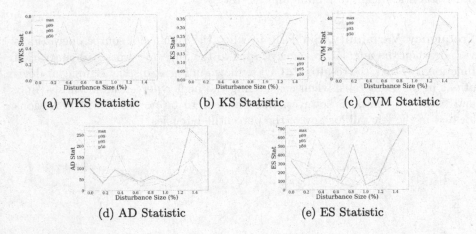

(a) WKS Statistic   (b) KS Statistic   (c) CVM Statistic

(d) AD Statistic   (e) ES Statistic

Fig. 7. Robustness of system to variability in amplitude

Amplitude Variability. We found the correlation between the WKS statistic and other (non-KS) statistics to be significantly lower for the higher percentile latencies when the amplitude of the Sinewave workload generation function is varied, compared to when the number of source operators is varied. This appears to manifest due to a larger proportion of the KS statistic values being located to the right of the median of the ECDFs, shown in Fig. 6, when the amplitude is varied, therefore being subject to the weighting function. In this instance, in general, the distributions differ most to the right of the median, albeit not to an extreme level. The weighting function applied to the KS statistic, described in Eq. 3, is non-linear so reduces the levels of correlation once applied.

Figure 7 shows that when the amplitude is varied, the CVM, AD and ES, shown in Figs. 7c, 7d and 7e respectively, exhibit large increases in the test statistic value for the higher percentile latencies (max and 99th) as the disturbance size reaches the maximum test range. This may signify a disturbance which causes the system to operate near the limit of its stable operating environment.

Sentence Size Variability. When the sentence size is varied, we observe the resulting ECDFs of the percentile latencies as displaying long right tails, with significant deviations within. Similarly to when the frequency input value is varied, the ECDFs show minimal deviations in the left tail. In this instance, it is noticeable that there also appear significant deviations between the ECDFs not only within the right tails, but within the main bodies of the distributions, albeit occurring to the right of the median. This large number of occurrences of maximum deviations to the right of the median has the same effect on the correlations between the WKS and other (non-KS) statistics, as seen for the higher percentile latencies when varying amplitude, as shown in Table 3.

Figure 8 appears to suggest a far lower level of system robustness across all the test statistics, and across all percentile latencies, when the workload sentence size is varied, as when compared to the case whereby the number of workload source operators is varied. All statistics display a non-linear relationship between the size of the change in test statistic value and the size of disturbance.

Combined Insight. From these experiments, we have seen how a single metric such as the KS statistic, whether weighted or not, may struggle to properly convey as much insight regarding a streaming system's level of robustness as could be conveyed through the application of multiple goodness-of-fit test statistics. When concerned with less extreme values and lower percentile latencies, the selection of test statistics presented here show strong overall similarities in their behaviour and in their interpretation of the system robustness. However, as the values under scrutiny become more extreme, whether from being contained within a higher percentile latency distribution, or by being located further towards the right tail of that distribution, the test statistics presented begin to diverge in their levels and their relative behaviours.

Table 3. Test stat correlation across latency %tiles w. variable sentence size

Statistic	Max					p99					p95					p50				
	WKS	KS	AD	ES	CVM	WKS	KS	AD	ES	CVM	WKS	KS	AD	ES	CVM	WKS	KS	AD	ES	CVM
WKS	1.0	0.78	0.43	0.76	0.44	1.0	0.87	0.65	0.9	0.63	1.0	0.89	0.81	0.9	0.76	1.0	0.95	0.93	0.86	0.91
KS		1.0	0.81	0.79	0.8		1.0	0.84	0.84	0.82		1.0	0.93	0.8	0.92		1.0	0.93	0.88	0.92
AD			1.0	0.74	0.99			1.0	0.74	0.99			1.0	0.83	0.99			1.0	0.97	0.99
ES				1.0	0.7				1.0	0.73				1.0	0.76				1.0	0.97
CVM					1.0					1.0					1.0					1.0

This should be viewed positively; the different stats tend to differ more when faced with more extreme values within the distribution, which can have greater importance. For example, while the Anderson Darling statistic seemed to show a lack of sensitivity to a situation with multiple ECDF cross overs occurring within the far right tail of a distribution, the ES statistic showed clear indications of a suggested lack of robustness when faced with the same set of inputs.

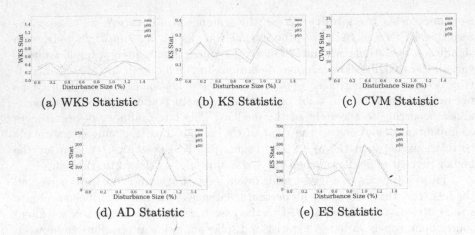

Fig. 8. Robustness of system to variability in sentence size

6   Conclusion

In this article we applied several non-parametric goodness-of-fit statistical tests to the measurement and quantification of streaming system robustness. We have shown that different goodness-of-fit tests produce differing relative measures of implied system robustness in the face of various disturbances to the incoming workload characteristics and workload function input variables.

The selected test statistics display differing relationships, interactions and levels of correlation between each other. Our results imply these are affected by not only the test statistics' inherent characteristics, sensitivities and calculation methods, but also by the particular percentile of observed latency values under scrutiny. In general, the higher the occurrence of observed latency values within the extreme tails of the distributions, or the higher the latency percentile under scrutiny, the larger the divergence between test statistics' results interpretation and the lower the levels of inter-statistic correlation.

This paper suggests that when seeking to quantify the robustness of distributed systems, practitioners should look to build multiple non-parametric goodness-of-fit measures into their analysis, rather than rely on a single metric.

We go on to suggest and outline areas of potential future research concerning goodness-of-fit tests based on a divergence measure, and entropy concepts. Further work could explore goodness of fit measures based on a "divergence measure", such as Kullback-Leibler [17] or Jeffery's divergence [13] which are based on entropy concepts. Relative entropy (or Kullback-Leibler divergence) is a measure of discrepancy between two probability distributions. It can serve as a measure of goodness of fit of any distribution to the other. Further work could entail investigating the appropriateness of employing entropy based goodness of fit tests in the domain of streaming system robustness measurement.

References

1. NEXMark benchmark. https://datalab.cs.pdx.edu/niagaraST/NEXMark/
2. Anderson, T.W., Darling, D.A.: Asymptotic theory of certain "goodness of fit" criteria based on stochastic processes. Ann. Math. Stat. **23**(2), 193–212 (1952). https://doi.org/10.1214/aoms/1177729437
3. Carbone, P., Katsifodimos, A., Ewen, S., Markl, V., Haridi, S., Tzoumas, K.: Apache flink^{TM}: stream and batch processing in a single engine. IEEE Data Eng. Bull. **38** (2015)
4. Carlson, J.M., Doyle, J.: Highly optimized tolerance: robustness and design in complex systems. Phys. Rev. Lett. **84**(11), 2529–2532 (2000). https://doi.org/10.1103/PhysRevLett.84.2529
5. England, D., Weissman, J., Sadagopan, J.: A new metric for robustness with application to job scheduling. In: HPDC-14. 2005 Proceedings of the 14th IEEE International Symposium on High Performance Distributed Computing, pp. 135–143. IEEE (2005). https://doi.org/10.1109/HPDC.2005.1520948
6. Engmann, S., Cousineau, D.: Comparing distributions: the two-sample Anderson-Darling test as an alternative to the Kolmogorov-Smirnov test. J. Appl. Quant. Methods **6**, 1–17 (2011)
7. Epps, T., Singleton, K.J.: An omnibus test for the two-sample problem using the empirical characteristic function. J. Stat. Comput. Simul. **26**(3–4), 177–203 (1986)
8. Frey, D.D., Engelhardt, F., Greitzer, E.M.: A role for "one-factor-at-a-time" experimentation in parameter design. Res. Eng. Design **14**(2), 65–74 (2003). https://doi.org/10.1007/s00163-002-0026-9
9. Goerg, S.J., Kaiser, J.: Nonparametric testing of distributions–the Epps-singleton two-sample test using the empirical characteristic function. Stata J. Promot. Commun. Stat. Stata **9**(3), 454–465 (2009). https://doi.org/10.1177/1536867X0900900307
10. Gribble, S.: Robustness in complex systems. In: Proceedings Eighth Workshop on Hot Topics in Operating Systems, pp. 21–26. IEEE Computer Society (2001). https://doi.org/10.1109/HOTOS.2001.990056
11. Harchol-Balter, M., Downey, A.: Exploiting process lifetime distributions for dynamic load balancing. ACM Trans. Comput. Syst. **15**, 253–285 (1997). https://doi.org/10.1145/224056.225838
12. Jamieson, S.: Dynamic scaling of distributed data-flows under uncertainty. In: Proceedings of the 14th ACM International Conference on Distributed and Event-Based Systems, DEBS 2020, pp. 230–233 (2020)
13. Jeffreys, H.S.: An invariant form for the prior probability in estimation problems. Proc. Roy. Soc. London Ser. A Math. Phys. Sci. **186**, 453–461 (1946)
14. Jen, E.: Stable or robust? What's the difference? CompLex **8**, 12–18 (2003)
15. Jensen, M.T.: Improving robustness and flexibility of tardiness and total flow-time job shops using robustness measures. Appl. Soft Comput. **1**(1), 35–52 (2001). https://doi.org/10.1016/S1568-4946(01)00005-9
16. Jorge Leon, V., David, W.S., Storer, R.H.: Robustness measures and robust scheduling for job shops. IIE Trans. **26**(5), 32–43 (1994). https://doi.org/10.1080/07408179408966626
17. Kullback, S., Leibler, R.A.: On information and sufficiency. Ann. Math. Stat. **22**(1), 79–86 (1951). https://doi.org/10.1214/aoms/1177729694
18. Linden, G.: Make data useful, slides from presentation at Stanford University data mining class (CS345) (2006). https://glinden.blogspot.com/2006/12/slides-from-my-talk-at-stanford.html

19. Linden, G.: Marissa Mayer at Web 2.0 (2006). https://glinden.blogspot.com/2006/11/marissa-mayer-at-web-20.html
20. von Mises, R.: Wahrscheinlichkeit Statistik und Wahrheit. Springer, Heidelberg (1928). https://doi.org/10.1007/978-3-662-36230-3
21. Mohamed, S., Forshaw, M., Thomas, N., Dinn, A.: Performance and dependability evaluation of distributed event-based systems: a dynamic code-injection approach. In: Proceedings of the 8th ACM/SPEC on International Conference on Performance Engineering, pp. 349–352 (2017)
22. Smirnov, N.V.: On the estimation of the discrepancy between empirical curves of distribution for two independent samples. Bull. Math. Univ. Moscou **2**(2), 3–14 (1939)
23. Tucker, P.A., Tufte, K., Papadimos, V., Maier, D.: NEXMark - a benchmark for queries over data streams draft (2002)
24. Wang, C., et al.: Performance troubleshooting in data centers: an annotated bibliography? ACM SIGOPS Oper. Syst. Rev. **47**(3), 50–62 (2013)

Effectiveness of Pre-computed Knowledge in Self-adaptation - A Robustness Study

Max Korn[✉][iD], Philipp Chrszon[iD], Sascha Klüppelholz[iD], Christel Baier[iD], and Sascha Wunderlich[iD]

Institute of Theoretical Computer Science, Technische Universität Dresden, Dresden, Germany
Max.Korn@tu-dresden.de

Abstract. Within classical MAPE-K control-loop structures for adaptive systems, knowledge gathered from monitoring the system and its environment is used to guide adaptation decisions at runtime. There are several approaches to enrich this knowledge base to improve the planning of adaptations. We consider a method where probabilistic model checking (PMC) is used at design time to compute results for various short-term objectives, such as the expected energy consumption, expected throughput, or probability of success. The variety PMC-results yield the basis for defining decision policies (PMC-based strategies) that operate at runtime and serve as heuristics to optimize for a given long-term objective. The main goal is to apply a robust decision making method that can deal with different kinds of uncertainty at runtime. In this paper, we thoroughly examine, quantify, and evaluate the potential of this approach with the help of an experimental study on an adaptive hardware platform, where the global objective addresses the trade-off between energy consumption and performance. The focus of this study is on the robustness of PMC-based strategies and their ability to dynamically manage situations, where the system at runtime operates under conditions that deviate from the (idealized) assumptions made in the preceding offline analysis.

1 Introduction

The world is full of dynamic systems, which adapt their behavior depending on the prevailing conditions with the simple goal to survive or to self-optimize with respect to various criteria, e.g., energy-efficiency. Nature already provides very efficient and smart ways for the autonomous management of adaption within biological systems and chemical processes. Likewise, there are human-created technical systems that also allow for adaptation to changing conditions either within

The authors are supported by the DFG through the TRR 248 (see https://perspicuous-computing.science, project ID 389792660), the Cluster of Excellence EXC 2050/1 (CeTI, project ID 390696704, as part of Germany's Excellence Strategy), and the Research Training Groups QuantLA (GRK 1763) and RoSI (GRK 1907) and the DFG-project BA-1679/11-1.

K. Gilly and N. Thomas (Eds.): EPEW 2022, LNCS 13659, pp. 19–34, 2023.
https://doi.org/10.1007/978-3-031-25049-1_2

the system itself or to external changes in the environment. In particular, there is a growing importance and dependence of complex dynamic hardware/software systems, systems of systems, and cyber-physical systems that are subject to different types of changes. Recent application areas include, e.g., autonomous driving and flying, where internal and external changes can happen anytime. Internal changes are, e.g., software updates or failures of hardware/software components. External or environmental changes include, e.g., bad road or weather conditions and sudden pedestrian appearance. Adaptations may also be required in case of a shift in the prioritization of different objectives or the rebalancing of trade-offs, e.g., performance vs. energy efficiency. Typical decisions for system adaptation include, e.g., switching on and off of components, switching between management policies, (de)allocating resources, and task scheduling and migration.

The challenge of maintaining and managing complex systems efficiently at runtime often exceeds what can easily be achieved by human beings. In practice, decision making is usually based on heuristics that generally do not provide optimal or close-to optimal results. In case a multitude of demands must be met, e.g., safety, reliability, throughput, and minimizing operating costs, decision making becomes increasingly complex. A recent trend is to direct (parts of) this complex task to learning-based approaches, which may be problematic in safety-critical settings where black-box decision making does not provide insights and hence trust. Alternatively, approaches based on formal methods can be employed to enable both transparent and (close-to) optimal decision making. In this paper, we apply probabilistic model checking (PMC) for supporting the runtime decision-making. The focus will be on the robustness of PMC-supported decision making when the stochastic assumptions on the environment behavior are inaccurate or biased and the system is subject to unforeseen dynamics.

The Approach. The general goal is to optimize runtime decision making in adaptive systems regarding a given *long-term objective*, which is typically a multi-objective property addressing the trade-off between conflicting objective functions. A typical example for such tradeoff is the minimization of energy consumption combined with the maximization of performance or throughput of a system. The goal is then to balance the tradeoff appropriately according to the application specific demands. Unfortunately in practice, the assumption that a detailed system model can be built and optimal solutions regarding a given long-term objective can be computed in time is often not valid. To anyhow achieve this goal, PMC-results for various *short-term objectives* with fixed horizon (such as step-bounded or reward-bounded objectives) are computed at design time. In this step, we use an abstract and discrete-time operational system model (i.e., a Markov decision process (MDP) annotated with costs/rewards) and store PMC-results into a database that can later be queried at runtime. The decision making can adapt the system according to adaptation choices that seem optimal in regards to selected short-term objectives and in a way that potentially approximates optimal solutions for the long-term objective. So called *PMC-based strategies* are query execution plans for the database that specify how exactly PMC-results for short-term results are combined within the decision making at runtime. In this paper we propose different PMC-based strategies and quantify the quality and robustness of PMC-based decision making. For this, we identified the following critical aspects to be addressed within the experimental studies: (A1) reducing the accuracy of the stochastic environment assumptions, as prior knowledge on the concrete environment behavior at runtime is in general not available, (A2) limiting the analysis horizon, as analyzing the system model to its full depth is practically impossible, and (A3) increasing the time between adaptations, as adaptation in each time step is unrealistic also due to the imposed adaptation costs, e.g. in terms of delays and additional energy.

Contributions. The main contribution of this paper is hence a **robustness study for PMC-supported decision making**. As a starting point we developed a stochastic operational model of a database application that runs on an adaptive hardware platform with alternative computational modes.

1. We suggest different PMC-based strategies for managing the trade-off between the energy consumption and performance (i.e., the amount of database queries that can be processed by the platform within a fixed timeframe),
2. extend PRISM's [16] simulator with features to enable an evaluation of PMC-based strategies using statistical model checking,
3. carry out an extensive, simulation-based evaluation of PMC-based adaptation strategies to quantify their quality and robustness regarding aspects (A1–A3) compared to a standard heuristic for resource management, and
4. provide general guidelines on how to deploy PMC-based decision making for robust operation depending on the critical application characteristics.

An artifact is made available under https://tud.link/8036, together with an extended version of this paper. The artifact contains the operational model, the full tooling, a demo example and the documentation. Furthermore, it allows for reproducing all experimental results.

Related Work. There is a large variety of formal methods addressing different components of the classical MAPE-K loop [13,21], namely for monitoring, analyzing, planning, executing and to gain additional knowledge. Existing approaches are closely related, but do not share the primary goal of robustness decision making. Instead, they are orthogonal and can be combined with PMC-supported decision making. Runtime verification (RV) is for example used for the monitoring of certain properties and their verification at runtime. RV is typically used for safety, reliability, security, fault containment, and recovery as well as for online system repair, but also for aiming at quantitative objectives [17]. Usually, RV does not rely on an abstract system model and model checking at runtime. In contrast, *online model checking* performs classical model checking periodically. The goal is to obtain guarantees on the near future in particular with respect to reliability (often covering safety only) [11,23], even though the model might be inaccurate or of approximate nature only. The limits of any approach that carries out model checking at runtime are determined by the time available for decision making, which is typically very short. *Incremental model checking* [14,18] hence considers an operational model for some base behavior plus models for dynamic aspects or components. This incremental model structure potentially allows for reducing the effort of model checking at runtime, as only the dynamic system parts have to be considered and combined with results from previous verification runs. The development of approaches for PMC-supported decision making and planning in adaptive software is an emerging research field [4,6,9]. Although there has been some progress on online approaches that call a probabilistic model checker for (bounded) models at runtime [5,19,20,22], PMC is too compute-intensive for a pure on-the-fly approach when decisions cannot afford delays and have to be made almost instantaneously. Incremental approaches (e.g. [10]) try to narrow this gap in performance, but limits on what can be computed at real time while the systems executes (or waits for adaptation) still exist. Hence, there is a clear motivation to carry out any kind of complex PMC analysis at design time rather than at runtime. The work in [8] also relies on a pre-analysis of quantitative measures (only) for reliability. The results are represented and stored as symbolic expressions, namely polynomials with free parameters that are then evaluated at runtime for the actual parameter values. The authors rely on parametric model-checking techniques. The clear benefit of this is, that the model checking results can be stored very compactly (in terms of symbolic expressions), but the approach is only practical when the number of free parameters is reasonably small. The method in [12], which we rely and build on in this paper, uses pre-analyzed quantitative measures to create annotated MDPs for runtime decision making. In contrast to [8], results are stored explicitly rather than symbolically. Our approach is similar in this aspect. Technically, we rely on a database to store results for multiple state properties. The database is then dynamically queried at runtime. Nevertheless, the common assumptions of [8] and [12] are, that the measures computed in the offline analysis are fixed and the exact same as the ones addressed and optimized for at runtime and that the full MDP model is small enough to be analyzed in its full depth. In our work, we drop those assumptions and address the question of how (for large systems in particular) short-term objectives and decision policies with bounded scope on parts or fractions of the system (which can be computed at design time) can be utilized and dynamically combined into complex adaptation policies (called PMC-based strategies) to optimize for a given long-term objective. The main question we address is then the aspect of robustness of different, alternative PMC-based strategies. We also want to stress the fact, that the selected method is compatible with other (formal) methods present in a MAPE-K loop. For instance, RV can be used for monitoring and online model checking can be used periodically and even incremental to update the database in the background.

2   PMC-Supported Decision Making

The classical MAPE-K control loop consists of a managed system operating within an uncontrollable environment, monitoring components (M) collecting information on the system and the environment via sensing at runtime, components to analyze the available information (A), components that generate an

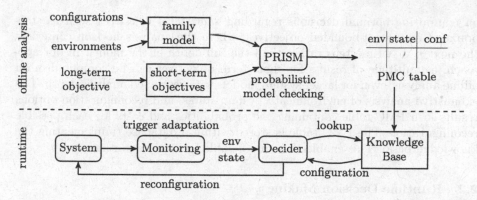

Fig. 1. Schema of PMC-supported decision making

adaptation plan (P) from the output of the analysis components as well as the knowledge base (K), and components that execute the current plan (E) via actuators on the system under control. In this work, the system under control is typically a hardware/software system whereas the environment is defined by the workload assigned to the system. In the following, we outline a PMC-based planning approach for steering the system adaptation as seen in Fig. 1. This approach requires an offline pre-analysis of the system for producing the knowledge base (tech. a database) used at runtime.

2.1 Model-Based Offline Analysis

Starting point of our procedure is an abstract and discrete-time operational model of the system in question (i.e., a Markov decision process (MDP) annotated with costs/rewards), which combines non-deterministic choices with probability distributions over successor states. The non-deterministic choices encode possible reconfiguration actions and the probability distributions encode stochastic assumptions, e.g., on the number of newly arriving tasks. Cost and reward structures are used to annotate, e.g. energy consumption in the given configuration. The PMC-analysis then takes the model together with a property specification (i.e., in terms of a PCTL* formula (probabilistic computational tree logic)) and computes so called *schedulers*, which resolve the non-deterministic choices in order to maximize (or minimize) expected accumulated costs as well as probabilities of temporal events[1]. In order to optimize decisions regarding a *long-term objective*, which is typically a multi-objective property, the goal is to balance the tradeoff in a way that fulfills the application specific demands. In [12], the MDP model is annotated with the results (per state) which then yields the basis for the decision making at runtime, on how to best achieve the targeted long-term objective. Since this is not always practical, we focus instead

---

[1] See [3] for details on MDPs, schedulers, and PCTL*-model checking.

on computing optimal decisions regarding *short-term objectives* such as step-bounded or reward-bounded objectives, where the analysis horizon comprises the next $n \in \mathbb{N}$ timesteps rather than the full depth of the model. In our approach, a multitude of results for short term objectives can be computed in an offline analysis, even for larger system models with restricted analysis depth. The exhaustive analysis of environments, system states, and reconfiguration options results in a PMC-table containing the probabilities and costs for each possible reconfiguration. The PMC-table is stored within a database (implementing the knowledge base). This enables fast and flexible lookups at runtime.

2.2   Runtime Decision Making

At system runtime, it is now possible to dynamically access the precomputed results and search for a reconfiguration action that is most promising for a given temporal objective. We call such a policy that is formulated as query on precomputed PMC-results a *PMC-based strategy*. Technically PMC-based strategies can be implemented as database queries: given a temporal objective Φ, a certain system state, assumptions on the environment, and optionally additional knowledge, context information and environment predictions one can define database queries looking for a reconfiguration policy that tries to optimize either Φ directly (if the results for Φ have been precomputed and stored for a matching situation into the database) or heuristically by combining other precomputed results. Applying such a PMC-based strategy at runtime will cause the resolution of all nondeterministic choices within the MDP (which yields a MC) and can be understood as online construction of an MDP scheduler. As this querying happens at runtime, one can customize and modify PMC-based strategies dynamically according to the current situation, predicted evolution, context information and most importantly the long-term temporal objective. With all the results from the pre-analysis and the ability of dynamically modifying the current reconfiguration according to changes, the hope is to steer the adaptive HW/SW systems very efficiently, outperforming, e.g., related techniques that do not expose such features and in particular simple heuristics such as greedy-based strategies that are currently used due to their simplicity.

3   Robustness of PMC-Supported Decision Making

In this section, we investigate the robustness of PMC-supported decision making. In particular, the approach is evaluated to answer the following research questions.

(RQ1) How can the trade-off between conflicting objectives be resolved?
(RQ2) How robust is the decision making in case of inaccurate assumptions?
(RQ3) What impact has the analysis horizon on the quality of adaptations?
(RQ4) What is the influence of different monitoring approaches?
(RQ5) What are general guidelines for configuring a PMC-based control loop?

To address (**RQ1**)–(**RQ5**), we examine an adaptive hardware/software system, whose abstract operational model generalizes to a larger class of producer-consumer systems. The PMC-supported decision making is evaluated using model-based simulation.

Adaptive System Example. We have developed an abstract stochastic operational model (an MDP) for an adaptive database system [15] that processes incoming computational tasks (i.e. queries) over the course of a day. The number of incoming tasks[2] is stochastically distributed and depends on the time of day. Incoming tasks are enqueued into an input buffer for delayed execution. The size of the task buffer is limited and an overflow results in dropped tasks, which in turn incurs an SLA-violation. The platform can be operated in various configurations regarding: the numbers of CPUs used, their frequency levels, and the usage of hyper-threading. Each configuration allows for processing a specific number of tasks per time step while consuming a certain amount of energy. The platform can be reconfigured in reaction to changing workloads or energy constraints. Our parametrized operational model can be instantiated with different resolutions of the time domain and e.g., different distributions for incoming tasks, buffer sizes, configurations and their cost/utility characteristics. Beyond this and due to the selected level of abstraction, the operational model generalizes to a class of producer-consumer systems with alternative operational modes. The specific values used throughout our experiments are based on the ones in [15]. The resolution of the time domain was chosen such that a large number of experiments can still be carried out in a reasonable time.

Long-Term Objective. Our long-term objective takes reference to the energy/utility tradeoff and are based on the two following measures[3] that refer to the probability of successfully processing tasks and the expected energy consumption.

$$\Pr(\text{Success}) \overset{\text{def}}{=} \Pr(\text{"no sla vio" U ("day end"} \wedge \text{"buffer empty"})) \quad (1)$$

$$\mathbb{E}(\text{Energy}) \overset{\text{def}}{=} \mathbb{E}_{\text{Energy}}(\lozenge\text{"day end"}) \quad (2)$$

Here, (1) is the probability of finishing all tasks by the end of the day without SLA-violations, i.e., the *success probability* (which needs to be maximized), and (2) quantifies the expected accumulated energy costs (which should be minimized).

PMC-Based Strategies. Obviously, the two measures are not independent and even conflicting. Maximizing the probability that all tasks are processed can be trivially achieved by running the system in the configuration with the highest possible throughput, but this also implies maximal energy consumption. Conversely, minimizing the energy consumption means selecting the configuration with the

---

[2] This abstract number and the respective distributions could also be used to characterize the computational weight of larger tasks, e.g., by the number of its subtasks.
[3] Alternative measures for utility could for example address latency.

least throughput or even turning off the system, resulting in multiple SLA-violations. To resolve this conflict, we formulate PMC-based strategies, which are gradually applied at runtime (i.e., a PMC-based strategy $\mathfrak{S}$ then induces a Markov chain $\mathcal{M}_{\mathfrak{S}}$). The ability of $\mathfrak{S}$ w.r.t. balancing the energy/utility tradeoff can be judged and quantified and with the help of $\mathrm{Pr}(\text{Success})$ and $\mathbb{E}(\text{Energy})$ evaluated in $\mathcal{M}_{\mathfrak{S}}$. The strategy $\mathfrak{S}_{\text{Util}\geq P}$ prioritizes the first objective by considering only those configurations where the probability of success[4] (i.e. no SLA-violation occurs) is greater or equal to P. Among the remaining configurations, the one with the lowest energy consumption is selected. $\mathfrak{S}_{\text{Budget:B}}$ tries to maximize the probability success, but selects only among those configurations that stay on energy budget B. Lastly, strategy $\mathfrak{S}_{\text{Quant:P}}$ relies on a multi-objective property. The idea is to find the minimal energy consumption that keeps the probability of success above threshold P.

Short-Term Objectives. Since the analysis horizon of the pre-computation is usually limited, the PMC-based strategies are defined in terms of the following *short-term objectives* to approximate the long-term objective.

(SO1) $\mathbb{E}_{\text{Energy}}^{\min}(\lozenge \text{ "horizon end"})$
(SO2) $\mathrm{Pr}^{\max}(\text{"no sla vio" U ("horizon end"} \wedge \text{"buffer empty")})$
(SO3) $\mathrm{Pr}^{\max}(\text{"no sla vio" U}^{\text{Energy}\leq B} \text{ ("horizon end"} \wedge \text{"buffer empty")})$
(SO4)
$\min e : \mathrm{Pr}^{\max}(\text{"no sla vio" U}^{\text{Energy}\leq e} \text{ ("horizon end"} \wedge \text{"buffer empty")}) > P$

Here, "horizon end" is an atomic proposition holding in all states where the end of the analysis horizon is reached. Formulas (SO2) and (SO3) are unbounded and bounded versions of PCTL-Until formulas, standing for the maximum probability for reaching the end of the analysis horizon without SLA-violation. The parameter B represents the energy budget for a single day. Formula (SO4) stands for a quantile [2]: given a fixed probability bound $P \in [0, 1]$, the quantile value represents the minimum amount of energy needed such that the energy-bounded version of (S02) holds with at least probability P.

The pre-analysis yields a table containing the PMC results for these objectives, depending on the current system state and the chosen adaptation in that state. At runtime, these results are queried according to the chosen PMC-based strategy. The strategy $\mathfrak{S}_{\text{Util}}$ first maximizes (SO2) and then minimizes (SO1), strategy $\mathfrak{S}_{\text{Budget:B}}$ maximizes (SO3), and strategy $\mathfrak{S}_{\text{Quant:P}}$ minimizes (SO4).

3.1 Experiment Setup

In this section, we give a brief overview of modeling details and the simulation infrastructure that was utilized for the evaluation.

Offline Analysis. For the offline analysis step, we make use of a family of MDP models that contains components for the adaptive hardware/software platform

---

[4] We simply write $\mathfrak{S}_{\text{Util}}$ when the probability bound P equals 1.

itself, the set of stochastic environment assumptions, and additional monitoring components that track runtime information and may trigger adaptations. Technically we rely on the tool PROFEAT [7], which supports feature-based modeling of system families and a family-based analysis. For the latter, PROFEAT internally relies on PRISM, as the input language extends the PRISM modelling language and is automatically translated to standard PRISM. For the experiments presented in the following, the model consists of 10 abstract time steps, where each time step consists of three phases. The first phase is the *adjust system* phase, where there is a nondeterministic choice among all available reconfiguration actions in the MDP, leading to successor states with respective operation modes. In the succeeding *sample workload* phase, the workload (i.e., 1–6 tasks) is sampled according to a given normal distribution $D(timestep)$ with a given time-dependent mean and a fixed variance. The incoming tasks are stored within an input buffer of size 12. In the experiments, we used ten synthetic environment assumptions with different shapes and 100 randomly generated environment assumptions. The last phase is the *system operation*, which processes as much tasks as possible within the chosen system configuration[5]. The other components of the family, e.g. environments, can be turned on and off following a set of rules, so we can analyse this family under different environments and configurations, and with different monitors active. The results are written to a database, where the keys are triples of the form $\langle s, m, \alpha \rangle$. Here, s is a state in the MDP that is part of the *adjust system* phase, m stands for monitored state, and α for a reconfiguration action.

Simulation-Based Evaluation. Rather than executing PMC-based adaptation on the actual system, we follow a simulation-based approach in order to evaluate the PMC-supported decision making. We extended the PRISM simulator to enable the use of PMC-based strategies for resolving all nondeterministic choices at certain points which are determined by an instance-specific monitoring component. The current system state s and monitor states m are then used in a database query to find the "best" adaptation action α w.r.t. the given PMC-strategy. This infrastructure now allows for using the statistical model checking engine of PRISM to compute probabilities and expectations in the Markov chain that results from applying the PMC-strategy to the runtime MDP model. We employ this simulator to gauge the quality of the decision making in various scenarios.

For our experiments, we used the confidence interval method of PRISM, with a sample size of 3000 paths and an 0.01 confidence. This means, that in every following experiment, 99% of results were in $\mathrm{Pr(Success)} \pm 0.02$ and $\mathbb{E}(\text{Energy}) \pm 1.4$ respectively.

3.2   Evaluation of PMC-Based Strategies

In the first set of experiments, we compare different PMC-based strategies to illustrate the potential of PMC-supported decision making and we examine their

---

[5] The concrete model can be found on https://tud.link/8036.

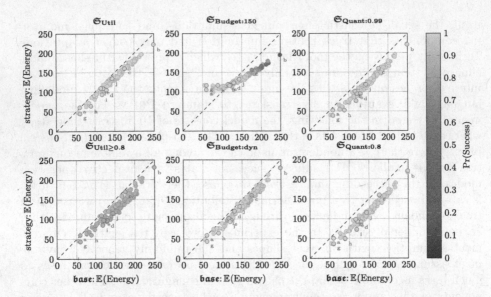

Fig. 2. Comparison: of baseline strategy **base** against PMC-based strategies

ability to balance long-term objectives (**RQ1**). The strategies are compared under idealized conditions, i.e., the runtime environment behavior agrees with the environment assumptions, the analysis horizon spans the whole day, and adaptations are triggered in every time step. As baseline for the overall adaptation quality, we consider a commonly applied heuristic (which we call **base**), which in each time step selects a configuration that uses the least amount of energy but has still high enough throughput to process all incoming tasks immediately. This heuristic has, e.g., been applied in the context of operating systems and the resource management of an adaptive database system [1,15]. We first show how much the PMC-strategies can further reduce the expected energy consumption. This is possible since the pre-analysis provides predictions which allow us to safely delay the processing of tasks without risking immediate SLA-violations.

Figure 2 shows the expected costs and the success probability of the different PMC-based strategies compared to **base** when applied to the 100 random and 10 synthetic environments with a variance of 2.0. The strategy $\mathfrak{S}_{Util}$ (or $\mathfrak{S}_{Util \geq 1.0}$) guarantees a success probability of 1.0 while reducing the expected costs. For the 100 random environments, the average energy saving was 11.8%. This makes $\mathfrak{S}_{Util}$ strictly more efficient than the baseline strategy. The expected energy costs can be lowered further by also lowering the bound for the success probability. The application of $\mathfrak{S}_{Util \geq 0.8}$ gains an average energy saving of 33.1%.

The $\mathfrak{S}_{Budget:B}$ strategy with a fixed energy budget B completely uses up the given budget even if less energy would have been needed to guarantee a high success probability. Thus, it is important to choose a budget that is close to the overall expected energy consumption for a certain environment. Applying this

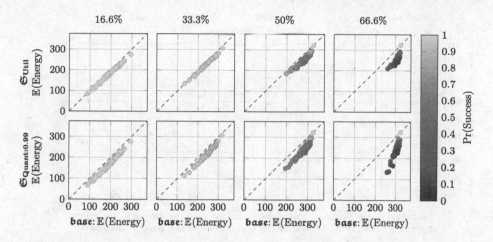

Fig. 3. Robustness of $\mathfrak{S}_{Util}$ and $\mathfrak{S}_{Quant:0.99}$ under noise (increased means)

idea yields the strategy $\mathfrak{S}_{Budget:dyn}$, where the budget is calculated for each individual environment and set to the expected energy consumption under the $base$ heuristic. This leads to an average cost improvement of 13.4% compared to $base$, while still providing a high success probability. The strategy $\mathfrak{S}_{Quant:P_{0.99}}$ is superior to both $\mathfrak{S}_{Util}$ and $\mathfrak{S}_{Budget:dyn}$, as it reduces the energy consumption by 18.5% on average while still providing a success probability of almost 1. Lowering the success probability bound to 0.8 leads to a further reduction of 21.88% compared to $base$. However, unlike the $\mathfrak{S}_{Util \geq P}$ strategy, lowering the probability bound P of strategy $\mathfrak{S}_{Quant:P}$ does not affect the success probability as much. In the setting with idealized assumptions, one can conclude that the considered PMC-based strategies can reliably outperform the baseline strategy $base$. In the following, the idealized assumptions are stepwise relaxed. Since the strategies $\mathfrak{S}_{Util}$ and $\mathfrak{S}_{Quant:0.99}$ are the most promising of the compared strategies, the following experiments will focus on these two strategies.

3.3 Impact of Unexpected Workloads

The focus will now be on the robustness of the PMC-based decision making when the actual load put onto the system is higher than assumed in the preceding offline analysis (**RQ2**). For each of the 110 environments, we gradually increase for each time step the means of the distributions $D(t)$ within the runtime environment by 16.6%, 33.3%, 50%, and up to 66.6% of the maximal workload processable in one time step. Figure 3 shows the effect on Pr(Success) and $\mathbb{E}$(Energy) for the PMC-based strategies $\mathfrak{S}_{Util}$ and $\mathfrak{S}_{Quant:0.99}$, again with baseline $base$.

The average energy consumption increases steadily, which is expected since increasing the overall number of tasks requires a higher throughput. The success probability decreases with an increasing unexpected load on the system. $\mathfrak{S}_{Util}$ is

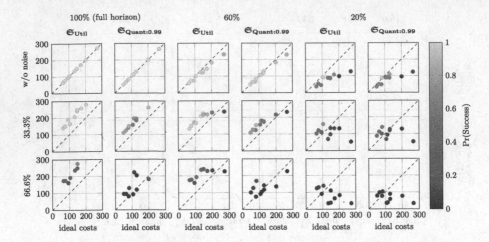

Fig. 4. Expected energy consumption and probability of success under increasing noise and decreasing analysis intervals compared to idealized conditions.

able to properly deal with an increase of up to 33.3%, without incurring higher costs than baseline ($\mathbb{E}$(Energy) is still 4.9% better compared to the baseline) and without significantly lower success probability (the average Pr(Success) is down to 0.94). From level 50% onward, $\mathfrak{S}_{Util}$ is no longer able to properly guide the system, since the difference between the expected number of arriving tasks and actually arriving tasks is simply too large. On the other hand, $\mathfrak{S}_{Quant:0.99}$ reacts slightly different to an unexpected load. Here, we also see a reasonable performance up to 33.3%, but with higher cost improvements of 11.3% at load level 33.3%. The average success probability drops down to 0.92, while the worst probability value is 0.79. For $\mathfrak{S}_{Quant:0.99}$ with load levels above 50%, costs drop drastically whenever the success probability is getting close to zero. This is due to the fact that $\mathfrak{S}_{Quant:0.99}$ optimizes for energy consumption only in case SLA-violations can be no longer avoided. From the above results we conclude that in the considered setting, both PMC-based strategies can reasonably tolerate noise levels of 30–40% of the maximum load processable in one time step.

3.4   Controllable Factors of the PMC-Supported Decision Making

Our remaining research questions concerned the configuration of the PMC-based control loop ((**RQ3**) and (**RQ4**)). While results under ideal conditions and under reasonable noise looked promising, the goal is to explore the approach under more realistic conditions. For this we carried out experiments, where (1) the analysis horizon does not cover the full day and (2) system reconfiguration is not happening in each timestep, but triggered by different monitors that observe the system and the environment. Results for (1) imply that analysis horizon should be above 40% and that (2) the monitor *env. change* serves best to minimize the number of reconfiguration steps (every second timestep) and maintaining the energy consumption and performance at a reasonable level. For the latter, the monitor *env. change* triggers adaption whenever the number of incoming tasks per timestep has changed by at least 2 since the last decision was made[6].

Insights from the individual experiments for (1) and (2) went into a combined experimental study (cf. Fig. 4) in which we investigated the robustness of the decision making under noise with limited analysis-horizon and using dynamic system reconfiguration (here using the monitor *env. change*).

Here, we see that both PMC-Strategies are still able to produce results of reasonable quality with analysis horizon higher than 40%, and a noise up to 33%. But it is also visible that the quality impacts add up, worsening the adaptation and creating some outliers, where short term goals can no longer be kept.

Finally, we want to answer (**RQ5**), and find some guidelines for choosing analysis horizon and monitoring. Generally, triggering adaptations in every time step yielded always the best results. However, should the cost of adaptation be not negligible, an environment-dependent monitor provides a suitable trade-off between the number of decisions and the adaptation quality. The choice of the PMC-based strategy depends on the overall goal, i.e., the long-term objective. As shown in Fig. 4, the $\mathfrak{S}_{Util}$ prioritizes the prevention of SLA violations at the cost of a higher energy consumption, while the $\mathfrak{S}_{Quant:0.99}$ allows for a greater reduction of energy costs at the risk of additional SLA violations. Finally, the analysis interval length should be chosen as short as possible to decrease the overhead of the pre-computation, but long enough to allow for the prediction of high-load situations. In our experiments, an analysis horizon higher then 40% was sufficient, independent from the noise level.

3.5   Performance and Scalability of PMC-Based Decision Making

All experiments have been carried out on a server with Intel(R) Xeon(R) E5-2680 CPUs with in total 16 physical cores and 377 GB of DDR3 RAM. For the quantitative analysis as well as for the decision making at runtime, we relied on a modified version of PRISM 4.5. For all experiments 8 GB of RAM were sufficient. For the database to store and look up PMC-results we relied on SQLite 3.30.1.

Offline Analysis. The base model used consists of 672 085 reachable states, the construction of which required less than one second using the Sparse engine of PRISM. The analysis times to compute the results for all 110 environments per property (SO1–SO4) depends on the considered horizon. Considering the maximal horizon, the total computation time (i.e., over all environments) for (S01) was still less than 1s, for (S02) it was less than 2 s, and for (S04) not more than 10 s. The budget considered in Formula (SO3) demands for an additional counter variable added to the model. This leads to an increased state space size of up to 67 618 570 states, when considering a budget of 125. Due to the increased state space, the total computation time for (S03) was longer, but was still less than 80 s. In most cases, the database consists of 3696 rows (i.e., possible

---

[6] Further information can be found in https://tud.link/8036.

configurations) and consumes about 1.3 MB of physical storage. Including (S03) and a monitor, this grew up to 6 054 048 rows (1.2 GB) of storage.

Runtime Evaluation. The evaluation at runtime relied on statistical model checking with 3000 simulation runs per experiment. Each run involves up to ten decision points where database lookups happen. Each lookup to the indexed database took about 1.1 ms on average when using (SO1, SO2, SO4) and about 15 ms for (SO3). In the largest setting the average lookup time grew up to 895 ms.

Threats to Validity. The proposed method suits well to adaptive producer-consumer systems whenever postponing tasks allows for operating the system in more energy efficient computational modes. Regarding scalability of PMC-supported decision making, it will be crucial that the database querying is fast enough. Reducing query-time was out of scope for this paper. On one hand the databases will easily grow and query time will increase when considering larger system models and/or PMC-based strategies involving complex system or environment monitors. On the other hand there is large room for improvements not yet considered for this article. By reducing the tables to only necessary information and using advanced database techniques, we are confident that even for larger system models, query times can be kept within reasonable bounds.

4   Conclusion

A goal in this paper was to show the general potential of PMC-supported decision making at runtime and in particular to study the robustness of the PMC-based strategies in practice. Our experiments have been carried out for an abstract version of an adaptive database server. As a first result, we were able to keep the time for database lookups and hence the overall delay for decision making reasonably low. The experiments demonstrate, that there exist robust PMC-based strategies, which operate with limited analysis horizon, under noise, and do at the same time, keep the number of reconfiguration steps (and hence cost for reconfiguration) within reasonable bounds. Given these promising results, a natural next step would be to apply the PMC-decision making to the actual adaptive system (rather than the simulation of the system). In this setting additional challenges are to be expected, and in particular with respect to delays and other concrete timing issues, which are due to the selected level of abstraction (regarding time) in the abstract system model as used in the offline-analysis.

References

1. Arega, F.M., Haehnel, M., Dargie, W.: Dynamic power management in a heterogeneous processor architecture. In: Knoop, J., Karl, W., Schulz, M., Inoue, K., Pionteck, T. (eds.) ARCS 2017. LNCS, vol. 10172, pp. 248–260. Springer, Cham (2017). https://doi.org/10.1007/978-3-319-54999-6_19
2. Baier, C., Daum, M., Dubslaff, C., Klein, J., Klüppelholz, S.: Energy-utility quantiles. In: Badger, J.M., Rozier, K.Y. (eds.) NFM 2014. LNCS, vol. 8430, pp. 285–299. Springer, Cham (2014). https://doi.org/10.1007/978-3-319-06200-6_24

3. Baier, C., Katoen, J.-P.: Principles of Model Checking. MIT Press, Cambridge (2008)
4. Calinescu, R., Ghezzi, C., Kwiatkowska, M.Z., Mirandola, R.: Self-adaptive software needs quantitative verification at runtime. Commun. ACM **55**(9), 69–77 (2012)
5. Calinescu, R., Grunske, L., Kwiatkowska, M.Z., Mirandola, R., Tamburrelli, G.: Dynamic QoS management and optimization in service-based systems. IEEE Trans. Softw. Eng. **37**(3), 387–409 (2011)
6. Cámara, J., de Lemos, R.: Evaluation of resilience in self-adaptive systems using probabilistic model-checking. In: 7th International Symposium on Software Engineering for Adaptive and Self-Managing Systems, SEAMS 2012, Zurich, Switzerland, 4–5 June 2012, pp. 53–62. IEEE (2012)
7. Chrszon, P., Dubslaff, C., Klüppelholz, S., Baier, C.: ProFeat: feature-oriented engineering for family-based probabilistic model checking. Formal Aspects Comput. **30**(1), 45–75 (2018)
8. Filieri, A., Ghezzi, C., Tamburrelli, G.: Run-time efficient probabilistic model checking. In: Taylor, R.N., Gall, H.C., Medvidovic, N., (eds.) Proceedings of the 33rd International Conference on Software Engineering, ICSE 2011, Waikiki, Honolulu, HI, USA, 21–28 May 2011, pp. 341–350. ACM (2011)
9. Filieri, A., Tamburrelli, G., Ghezzi, C.: Supporting self-adaptation via quantitative verification and sensitivity analysis at run time. IEEE Trans. Softw. Eng. **42**(1), 75–99 (2016)
10. Forejt, V., Kwiatkowska, M., Parker, D., Qu, H., Ujma, M.: Incremental runtime verification of probabilistic systems. In: Qadeer, S., Tasiran, S. (eds.) RV 2012. LNCS, vol. 7687, pp. 314–319. Springer, Heidelberg (2013). https://doi.org/10.1007/978-3-642-35632-2_30
11. Güdemann, M., Ortmeier, F., Reif, W.: Safety and dependability analysis of self-adaptive systems. In: Second International Symposium on Leveraging Applications of Formal Methods, Verification and Validation (isola 2006), pp. 177–184 (2006)
12. Ghezzi, C., Pinto, L.S., Spoletini, P., Tamburrelli, G.: Managing non-functional uncertainty via model-driven adaptivity. In: Notkin, D., Cheng, B.H.C., Pohl, K., (eds.) 35th International Conference on Software Engineering, ICSE 2013, San Francisco, CA, USA, 18–26 May 2013, pp. 33–42. IEEE (2013)
13. Hachicha, M., Halima, R.B., Kacem, A.H.: Formal verification approaches of self-adaptive systems: a survey. In: Knowledge-Based and Intelligent Information & Engineering Systems: Proceedings of the 23rd International Conference KES-2019, Budapest, Hungary, 4–6 September 2019, vol. 159 of Procedia Computer Science, pp. 1853–1862. Elsevier (2019)
14. Johnson, K., Calinescu, R., Kikuchi, S.: An incremental verification framework for component-based software systems. In: CBSE 2013, Proceedings of the 16th ACM SIGSOFT Symposium on Component Based Software Engineering, part of Comparch 2013, Vancouver, BC, Canada, 17–21 June 2013, pp. 33–42. ACM (2013)
15. Kissinger, T., Habich, D., Lehner, W.: Adaptive energy-control for in-memory database systems. In: Proceedings of the 2018 International Conference on Management of Data, SIGMOD Conference 2018, Houston, TX, USA, 10–15 June 2018, pp. 351–364. ACM (2018)
16. Kwiatkowska, M., Norman, G., Parker, D.: PRISM 4.0: verification of probabilistic real-time systems. In: Gopalakrishnan, G., Qadeer, S. (eds.) CAV 2011. LNCS, vol. 6806, pp. 585–591. Springer, Heidelberg (2011). https://doi.org/10.1007/978-3-642-22110-1_47

17. Leucker, M., Schallhart, C.: A brief account of runtime verification. J. Logic Algebraic Program. **78**(5), 293–303 (2009)
18. Lochau, M., Mennicke, S., Baller, H., Ribbeck, L.: Incremental model checking of delta-oriented software product lines. J. Logical Algebraic Methods Program. **85**(1), 245–267 (2016)
19. Nia, M.A., Kargahi, M., Faghih, F.: Probabilistic approximation of runtime quantitative verification in self-adaptive systems. Microprocess. Microsyst. **72**, 102943 (2020)
20. Rinast, J.: An online model-checking framework for timed automata. PhD thesis, Hamburg University of Technology (2015)
21. Weyns, D., Iftikhar, M.U., De La Iglesia, D.G., Ahmad, T.: A survey of formal methods in self-adaptive systems. In: Fifth International C* Conference on Computer Science & Software Engineering, C3S2E 2012, Montreal, QC, Canada, 27–29 June 2012, pp. 67–79. ACM (2012)
22. Zhao, Y.: Online Model Checking Mechanisms and Its Applications. PhD thesis, Universität Paderborn (2016)
23. Zhao, Y., Rammig, F.: Online model checking for dependable real-time systems. In: 2012 IEEE 15th International Symposium on Object/Component/Service-Oriented Real-Time Distributed Computing, pp. 154–161 (2012). ISSN: 2375–5261

Applications

Energy Consumption Estimation for Electric Vehicles Using Routing API Data

Saad Alateef$^{(\boxtimes)}$ and Nigel Thomas

School of Computing Science, Newcastle University, Newcastle upon Tyne, UK
saad.alateef@ncl.ac.uk

Abstract. Electric vehicle (EV) range anxiety influences electric vehicles' low penetration into the transportation system. There have been several developments in range estimation for electric vehicles. However, the studies that focus on determining the remaining range based on real-time publicly available data remain low. Most of the current methods employed consider limited data collection and do not consider the most substantial factors that directly impact energy consumption. This paper introduces a velocity model based on route information for the range estimation of electric vehicles. It uses publicly available data sets from several map service APIs and incorporates them into the range estimation algorithm. Three map service APIs were used to collect the data over an extended period. Then we analysed this data to extract the most representative data to generate the velocity profiles. The paper uses MATLAB code and Python libraries to process the representative data and apply the velocity model. Moreover, we have integrated it into an electric vehicle model, including the battery, to estimate the power demand for each trip and the remaining driving range. We observed that producing realistic driving cycles using public data is possible; furthermore, it simulates the driving patterns and satisfies the constraints of the vehicle.

Keywords: Electric vehicles · Driving cycles · Range estimation · SOC estimation

1 Introduction

Electrifying transportation is one of the main targets for the transportation sector to reduce greenhouse emissions in most countries [1]. However, Internal Combustion Engines (ICEs) are entirely dependent upon fossil fuels and still the primary propulsion system in road transport globally. The increase in the dependency on oil is considered significant as a result [2]. Therefore, there is an essential need to overcome this issue to increase the sustainability of the transportation system and address the environmental issues. The demand for electric vehicles has been increasing recently in the transportation markets, and

© The Author(s), under exclusive license to Springer Nature Switzerland AG 2023
K. Gilly and N. Thomas (Eds.): EPEW 2022, LNCS 13659, pp. 37–53, 2023.
https://doi.org/10.1007/978-3-031-25049-1_3

it is expected to continue to replace traditional vehicles in the next few decades. Electric vehicles (EVs) are an intelligent solution for the planet and will reduce gas emissions significantly [3]. However, range anxiety is one of the main challenges that face electrifying transportation, and it affects the adoption of electric vehicles.

In addition to the enormous advantage of reducing the pollution levels of EVs, this invention has some other benefits over conventional vehicles. These benefits include energy recovery when the battery restores some of the energy due to braking and the noise-freeness [4]. Regenerative braking is a crucial characteristic of EVs when the generator returns the energy to the battery system due to braking. According to previous studies, this feature is practical, especially in city driving and the daily commute. However, it is less effective on the motorway, and long journeys [5]. In addition, conventional vehicles consume more energy in city driving because of the heat loss due to braking in contrast with EVs [3].

This paper aims to develop a velocity model using publicly available routing data on specific routes. It attempts to construct the speed profile for a specific journey between origin and destination using the map API. After generating the potential realistic driving profile, we used a generic EV model to generate the potential power demand for the trip. Hence we apply the state of charge estimation method to analyse the impact of the route and traffic on the battery efficiency. This research concentrates on developing a data collection process using multiple map service APIs. Many drivers rely on the GPS data provided by map services to navigate to their destinations [6]. This paper uses the data collected from the drivers using the map API. The first step of this paper involves exploring the routing information and using it to estimate energy consumption and improve the battery-powered vehicles' efficiency. This research explores the data of three different map information providers through their API. Google Maps API [7], HERE Maps API [8], and TomTom Maps API [9] are the primary data sources in this research.

The amount of data collected from vehicles and drivers can significantly improve the range of electric vehicles [10]. The battery management system (BMS) installed in electric vehicles senses the battery's state of charge. It predicts the remaining range based on the battery status and other data installed on the system, such as the vehicles' specifications. However, BSM uses the range values to estimate the EV's remaining range and does not consider route information ahead. Therefore, it uses the range values for its estimation. The proper use of the available data can improve driving range prediction and energy consumption estimation.

In this paper we construct near to real-time velocity profiles to allow us to generate power profiles and estimate the power consumption before performing the journey.

This paper is organised as follows: Sect. 2 presents the related work to this field including energy consumption estimation and driving cycles. Section 3 explores the data collection procedures and methodology. Section 4 illustrates the driving cycle production steps based on some representative driving profiles. Section 5 present the electric vehicle model used to generate the power demand and to estimate the energy consumption using the driving cycles produced. Section 6 presents and discussed some the model's results and finally, Sect. 7 concludes the research and highlighted the penitential future work.

2   Energy Consumption and Driving Cycles

Energy consumption in transportation systems has been a significant research and development topic recently [11]. Previous work focused on how driving behaviours affect fuel consumption in ICE vehicles [12]. In recent years, further studies have been conducted on the usage and consumption of EVs [13]. These studies are characterised based on their methodology and purpose [14]. In addition, some researchers focus on the energy models of electric vehicles to improve the EV design [15], exploring the influential factors on power consumption [16] and the influence of the driving patterns on energy consumption and the remaining driving range [17].

Although there are many studies in the literature to improve the energy consumption of electric vehicles, there needs to be more research conducted on energy consumption based on real-time velocity profile prediction. These profiles are known as driving cycles for vehicles and are generally defined as a series of points representing speed versus time. The driving cycle is usually performed as a physical journey on a vehicle for various purposes and based on various criteria [18]. Driving cycles developed in recent decades are a standard tool for estimating fuel consumption and measuring the levels of air pollution produced by the transportation system. Many existing standard industrial driving cycles, such as NYCC, UDDS, and HWFET, have been used in some studies [19,20]. These driving cycles are used as velocity profiles to validate the response of the EV and battery models. However, the current driving cycles are performed in unknown conditions and do not represent real-time driving conditions. Some existing studies developed methods to predict the driving profile [21,22], and each relies on the nature of the data used to develop this prediction method [23].

The map service API can help to develop and improve real-time driving cycle construction methods. The API provides a wide range of route information for any geographical location on the map and considers the traffic situation. Even though the API providers restrict developers from some features for commercial and competition reasons, it is still possible to extract some valuable data to help to predict the journey and the velocity characteristics to improve the range and energy estimation for electric vehicles. Furthermore, this approach makes it more convenient than performing the physical journey considering many arrangements and set-ups such as a vehicle, driver and some equipment, making it a costly task [24].

3   Data Collection Process and Analysis

3.1   Traffic Data Exploration

This section illustrates the process and the purpose of exploring the traffic data. In addition, the data collection process and the challenges faced are also presented.

1. Route Selection:
   The main objective of collecting the data from the map service providers is to create a generic script that gathers time-specific traffic data between two different geographical locations following a specific route. We have specified the origin and destination on the map for two different routes that have different road structures. These routes were sliced into multiple chunks so that we can collect more accurate data for each chunk. Collecting the data for smaller segments is to separate the parts of the route that have possibilities of speed reduction from more continuous high-speed such as motorways.

2. Data Analysis:
   The data provided from the APIs are "duration", "distance" and "segments". Each segment profile includes duration and distance. Since the distance and the time are known, we can calculate the average speed for each segment and therefore, for the entire route. The plots for these raw collected data gives us an idea of what the speed profile, as it presents the average speed for each segment of the route.

3. Data Manipulation:
   Since the data obtained from the APIs are only average speed based on the duration in traffic and distance of the segment, it provides a constant speed for each chunk of the road. Therefore, we introduce some changes to those average speeds to reflect more realistic driving patterns. Therefore, it can represent the velocity of the vehicle in each segment without altering the mean value of the speed provided from the API data

3.2   Data Collection

1. Data collection methodology.
   - Extracting the data from the API provider.
   - Collecting data from the API response.
   - Scheduling the collection process for specific times.
   - Loading the data into a CSV format.
2. Source of traffic data.
   - Google Maps API
     The API products provided from Google Maps were used as follows:
     • Distance Matrix API: This API allows us to get the travel distance and time for the entire route and each identified segment. In addition, it allows us to obtain the estimated duration within the current traffic.

- Directions API: Allows introducing the way-points which helps force the API to follow the route we specify; it is also responsible for the mode of transportation, which is "Car" in our case.
  - TomTom Maps API
    The API products provided from TomTom were used as follows:
    - Traffic Flow API: This allows developers to request the travel time from the origin and destination with respect to the real-time traffic.
    - Maps API: This product gives an access to the API data every time we make a request.
    - Routing API: This API gives highly detailed information about the route, with respect to directions and travel mode.
  - HERE Maps API
    The API products provided from HERE Maps were used as follows:
    - Routing API: This product informs the estimated arrival time between the origin and destination.
    - Traffic API: This API is responsible for reporting the traffic flow, its consequences and the incidents information.
    - Way-points sequence API: This allows us to specify the way-points on the route to divide it to the segments we require.

3. Extracting the time and speed data
   - The data of the time taken during current traffic and the average speed calculated are added into separate files for each journey. These files are formatted in two columns that show the time in seconds for the whole journey versus the average speed at each second. These files are then processed to generate possible velocity profiles (Table 1).

Table 1. The main features of the used map services API are illustrated

	Google map	HERE map	TomTom map
Free transactions	40000 requests per month	250000 requests per month	2500 requests per day
Pricing	$5 for up to 100,000 requests	$1 for 1000 requests	$0.5 for 1000 requests
Technology	Direction and distance matrix APIs are called. Response in JSON format	Routing API is called. Response in JSON format	Routing API is called. Response in JSON format
Way-points limit	23 way-point in each request	50 way-points in each request	No way-point limits

The data was collected at multiple time-slots for each API. These slots were at 8:15, 12:00, 16:45 and 00:00. This time selection was done to evaluate and analyse the differences between the peak traffic hours and when it is quiet.

During each slot, the data is requested for an hour, and then loaded the data into CSV files in several rows. The number of rows are dependent upon how many intermediate points were introduced. The data consists of many columns starting from the date when the data was collected, until the average speed that was calculated using the distance and the duration in traffic. Each row is a repetition of the same process during the specific time we selected. Figure 1 illustrates a step by-step-process of collecting the data through the APIs.

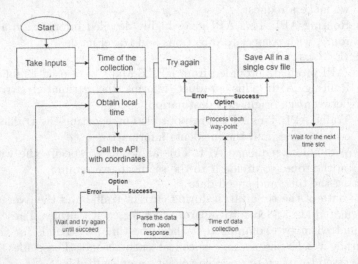

Fig. 1. Data collection process

4   Route Based Driving Cycle Construction

This section describes how we applied the acceleration and deceleration processes between the speed intervals, followed by the addition of the velocity variations to produce fluctuations in the speed profile wherever it remains constant. In addition, it illustrates the method used to smooth the velocity curves.

4.1   Applying Acceleration and Deceleration Between Route Segments

We applied the acceleration and deceleration rates to the beginning and ending intervals to smooth the velocity transition between segments. Based on Nissan Leaf's 2019 [25] acceleration rate for 0–100 km/h, we determine the maximum acceleration on the car [26]. In Eq. 1, the variables v_1 and v_2 are the speeds of two consecutive intervals. The speed variation Δ_v denotes if the vehicle is accelerating ($\Delta_v > 0, a >$) or decelerating ($\Delta_v < 0, a < 0$). If the speed remains constant between intervals ($\Delta_v = 0$), then the acceleration is zero.

We determine that for a speed variation that equals or exceeds $100\,\mathrm{km/h}$, the acceleration function saturates at its maximum value, $3.51\,\mathrm{m/s^2}$. In the case of deceleration, the saturation occurs at $-3.51\,\mathrm{m/s^2}$. For a variation between 0 and $100\,\mathrm{km/h}$, the acceleration function is linear, in which the coefficient multiplying Δ_v ensures the curve continuity at $\Delta_v = \pm100\,\mathrm{km/h}$.

$$a = \begin{cases} \frac{\Delta v}{28.49}\ \mathrm{m/s} & \text{if } |\Delta v| < 100\ \mathrm{km/h} \\ \pm3.51\ \mathrm{m/s} & \text{if } |\Delta v| \geq 100\ \mathrm{km/h} \end{cases} \quad \text{for } \Delta v = v_2 - v_1 \qquad (1)$$

The acceleration and deceleration intervals rely on the speeds from the previous $(i-1)$, the actual (i) and the next $(i+1)$ segments. The acceleration occurs at the beginning of the segment, while the deceleration occurs at the end. If none of the conditions below are satisfied, the speed along the segment is constant. Figure 2 explains the acceleration process in a function of speed variation.

$$\begin{cases} \text{if } v_i > v_{i-1} \text{ then accelerate } (a > 0) \\ \text{if } v_i > v_{i+1} \text{ then decelerate } (a < 0) \end{cases} \qquad (2)$$

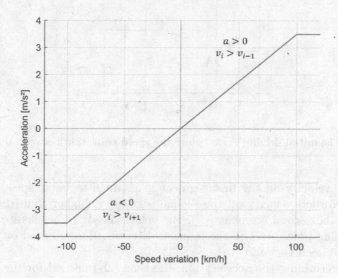

Fig. 2. Acceleration (a) in function of speed variation (Δv)

Using the data retrieved from the APIs, as shown in Fig. 3 we obtain the initial driving cycle. It is characterised by sharp edges, corresponding with unrealistic significant speed changes. In addition it does not take into account the technical constraints imposed by the vehicle and the road characteristics. Therefore, the final driving cycle needs to be developed realistically before performing the energy consumption estimation.

The process of developing the driving cycle is implemented in an iterative manners. In Fig. 4, the driving cycle shows three different segments with constant

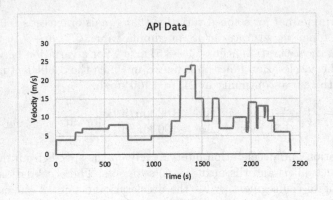

Fig. 3. Mean velocity obtained from HERE Maps API

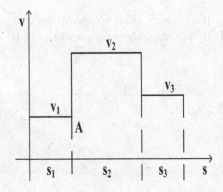

Fig. 4. The initial driving cycle before the speed transition between segments

speeds. The velocity on the first segment is assumed to be at speed (v_1), and since the recorded velocity on the second segment is higher than the vehicle's velocity on the second segment (v_2), the vehicle needs to accelerate gradually after exceeding point A. The acceleration determination is based on the speed difference between (v_1) and (v_2).

After determining the acceleration, the time (Δ_t) needed for the vehicle to accelerate from the velocity in the first segment (v_1) to the following velocity (v_2) can be calculated as:

$$\Delta t = \frac{v_2 - v_1}{a} \tag{3}$$

The distance (Δs) the vehicle needs during the accelerating process leads to the division of the following segment into separated segments as shown in Fig. 5. The Distance (Δs) can be calculated by:

$$\Delta s = v_1 \Delta t + \frac{a \Delta t^2}{2} \tag{4}$$

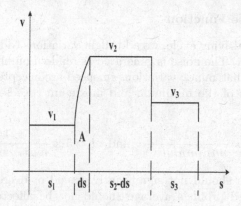

Fig. 5. The gradual acceleration added to the driving cycle

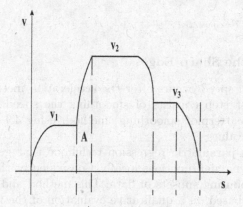

Fig. 6. Final driving cycle after applying the acceleration method

In addition, in Fig. 5, the first segment has a length of (ds) where the vehicle acceleration is applied until it reaches the speed of (v_2). The second segment has a length $s_2 - ds$ when the vehicle's velocity is constant and equals v_2. The API speed data is often imperfect and inconsistent; it deviates from real-life conditions and constraints. Therefore, the acceleration between velocities is not always feasible. In other words, in the above-analysed case of the acceleration from v_1 to v_2, sometimes the distance the vehicle needs to accelerate is longer than the length of the following segment. Therefore, to overcome this issue, the acceleration to v_2 will not take place; moreover, we reduce the speed on the following segment by a small step of Δ and repeat the process where the speed on the next segment is $v_2 - \Delta$. This process is repeated until it satisfies the feasibility of yielding the final driving cycle, as shown in Fig. 6.

4.2   Adding Noise Function

To mimic an actual driving cycle, we add slight variations to the intervals where the speed is constant. The noise is generated as random numbers in the interval (a, b). Considering that minor variations in speed are accepted, (a) and (b) are defined as functions of the minimum and maximum speeds of an interval (i), respectively.

$$a = -5 \times \frac{1}{minimum(v_i)} \quad \text{and} \quad b = 5 \times \frac{1}{maximum(v_i)} \tag{5}$$

The added variation should not significantly extend or decrease the route distance, and the speed profile's average should not be affected. Therefore, the mean of the noise must be zero. To ensure this condition, after the noise (n) is generated for (N) samples, it is corrected as follows.

$$n_{i \, corrected} = n_i - \overline{n}, \quad i = 1, 2, \cdots, N \tag{6}$$

4.3   Smoothing the Sharp Edges

A sharp variation in speed occurs after the acceleration method and the noise addition, so the last step consists of smoothing the speed curve-the LOESS (locally estimated scatterplot smoothing) method, using 4% of the samples to calculate smoothed values.

LOESS is a non-parametric regression technique that generates a smooth curve by fitting polynomial functions locally. Consequently, the fitted values are derived from neighbouring subsets of data. This method and the percentage of samples are chosen based on a qualitative evaluation of the final driving cycle. The primary requirements are the minimisation of sharp edges, the preservation of noise-induced changes, and the preservation of the cycle's original form. We establish that the cycle begins and finishes at 0 m/s, which is not always guaranteed in the smoothing function. The speed curve is linearly interpolated from zero to an arbitrary speed value at the beginning and end of each cycle. In addition, the speed in the first and final segments is usually low, and this method has been used to provide a smooth transition at the beginning and completion of each cycle.

We obtain the representative driving cycles illustrated in Fig. 7 after using the previous procedures. After developing the driving cycles for each route using data from the three various APIs, it is evident that the driving cycle for each API is diverse at certain parts along the route and relatively similar at others. The produced driving cycles will be utilised to construct the power profile for the electric vehicle in the next section. As a result, an energy estimate may well be conducted, and battery dynamics can be monitored.

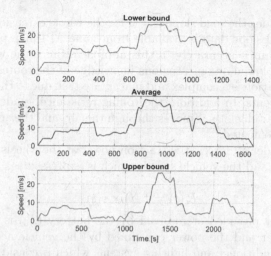

Fig. 7. Google Maps driving cycles for Route 1

5 Generating the Power Demand Using Electric Vehicle's Dynamics

This section consider an electric vehicle model based on existing Nissan Leaf to perform the power demand generation and the state of charge estimation based on the data used on this research. With the vehicle speed determined in the driving cycle, we calculate the power consumed to generate the vehicle, or, in case of braking, the power provided back to the battery pack [25] (Fig. 8).

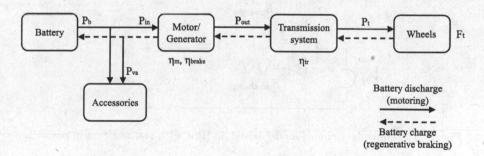

Fig. 8. Electric vehicle power transition diagram

$$F_t(t) = F_r(t) + F_g(t) + F_d(t) + F_a(t) \tag{7}$$

Starting at the wheels, the traction force (F_t) required for the vehicle's motion is expressed by the sum of opposing forces, which is the rolling friction (F_r), grade resistance (F_g), aerodynamic drag (F_d), and acceleration force (F_a) [27]

and [25]. We consider the road slope $\alpha = 0$ for the whole extension of the routes. Although the road slope data is available from some API map providers, it was impossible to obtain it accurately in this approach. Since the way-points were manually selected based on our previous knowledge of the routes, obtaining the road slope information is a complex and inaccurate task due to the uneven length of the route segments. In addition, the rolling resistance, grade resistance, the force opposing the vehicle as it moves through the air, and the acceleration force formulas are implemented.

After we consider all forces, the traction power at the wheels is a function of the traction force and the vehicle speed, which is expressed as:

$$P_t(t) = f_t(t) \times v(t) \tag{8}$$

Finally, the power provided or received by the battery (P_b) is the sum of the motor input power and the power consumed by the vehicle accessories (P_va), such as the air conditioner and lighting systems which is considered constant.

$$P_b(t) = P_{in}(t) + P_{va}(t) \tag{9}$$

5.1 Battery Model Dynamics and Energy Consumption Estimation

The Rint model, which was suggested in [28], was taken into account throughout the battery model's implementation. This model includes a voltage source V_{oc}, representing the open-circuit voltage, in series with the parallel branch of internal resistance. Any battery model can be implemented in this part of the research to estimate the state of charge based on our power profiles. The current model is less complex and validated in previous studies such as in [28] (Fig. 9).

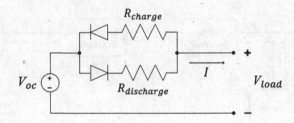

Fig. 9. The equivalent circuit model based on Rint with two resistors in parallel

The current flow in the resisting branch is represented by ideal diodes. When the battery is discharging, the diode in series with the discharging resistance ($R_{discharge}$) conducts the current; contrarily, in case of battery charge, the diode conducting the current is in series with charging resistance (R_{charge}). Given an initial state-of-charge, we start by calculating open-circuit voltage V_{oc} in terms of the state-of-charge (SOC), where K, a, b, c and d are constants that were taken from [25] and adjusted for a better response. This adjustment was empirical, and can be improved in the future.

$$V_{oc}(t) = K - a \times SOC(t) - b\frac{1}{SOC(t)}$$
$$+c \times \ln(SOC(t)) = d \times ln(1 - SOC(t)) \tag{10}$$

The charging or discharging resistance R_s is a function of the SOC and is determined based on look-up tables obtained from [25]. Then, the battery current is calculated by:

$$I(t) = \frac{V_{oc}(t) - \sqrt{V_{oc}(t)^2 - 4R_sP_b(t)}}{2R_s} \tag{11}$$

The current is positive if the battery is discharging, and negative if it is charging. Finally, the SOC is estimated with the coulomb counting method [29], in which the battery current is integrated over time to calculate the transferred charge.

$$SOC(t) = SOC(t_0) - \frac{1}{C_r}\int_{t_0}^{t} I\Delta\tau \tag{12}$$

where $SOC(t)$ is the current state-of-charge, $SOC(t_0)$ is the initial state-of-charge, C_r is the rated capacity, I is the current flowing in or out of the battery, t_0 is the initial time and t, the current time. Alternatively, the SOC can be expressed in terms of its previously estimated value $SOC(t-1)$ and the current for the time interval of $\Delta\tau = [t-1, t]$.

$$SOC(t) = SOC(t - 1) + \frac{I(t)\Delta\tau}{C_r} \tag{13}$$

The equations of V_{oc}, I and SOC are applied iteratively over time to obtain the profiles for a full driving cycle.

6 Results

This section presents the power needed for some journey based on each API and route. It also shows the battery voltage and the state of charge estimation.

Figure 10, shows the power profile, battery voltage and the state of charge for Route 2 based on the data obtained from Google Maps API. The start and the end of the route illustrate the impact of the speed variation due to the excessive acceleration and deceleration. In addition, restricted variations due to the heavy traffic in the middle of the motorway increased the energy consumption due to less battery recovery.

Figure 11, illustrates the same results based on HERE Maps API. The power profile has more variations compared to Google case, with more excessive acceleration along the journey. The final energy consumption is slightly similar to Google which could be as a result of the regenerative braking.

Figure 12, presents the results for the same journey based on TomTom Maps API. This shows the highest power demand due to the higher speed compared to the others. However, it has similar profile at the middle of the motorway compared to Google Maps. Moreover this journey shows more energy consumption due to the higher speed at some segments of the route and less battery recovery at the middle of the journey.

Time was adopted as the reference for the graphs at the (x-axis) and equations applied in the driving cycle generation, which is especially helpful in acceleration and deceleration intervals. In graphs, it highlights the influence of traffic on the duration of the trip.

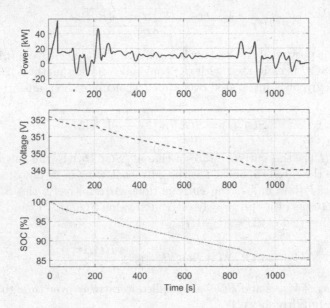

Fig. 10. Lower bound for Google Maps data Route 2

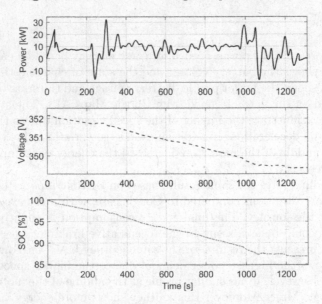

Fig. 11. Lower bound for HERE Maps data Route 2

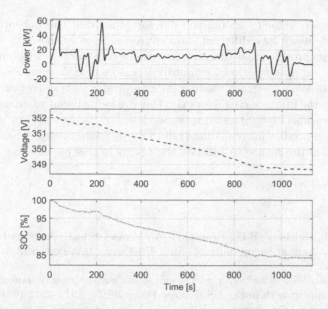

Fig. 12. Lower bound for TomTom Maps data Route 2

7 Conclusion and Future Work

This paper argued to construct different driving cycles based on three API data and two different routes. A data collection framework is developed which gathers the same data from different API and process the data to generate realistic driving cycles. We divided the routes into slices using the route segmentation technique. The data contain the distance of the journey, the time taken for the whole journey, the average speed for the whole journey, and the way-points. We developed a velocity model algorithm and introduced variations to the velocity profiles.

To smooth down any sharp edges in the randomised data, a locally weighted scatterplot smoothing function (LOWESS) was applied. The data selection is based on data classification and statistical analysis. An electric vehicle's model based on Nissan Leaf was implemented to calculate the power demand and the remaining range for each cycle.

The results show that the driving cycles are within the range of the existing industrial cycles, simulating realistic driving patterns for each cycle and the vehicle's constraints are satisfied. The results also show the variation between the different data sources and the times for the data collection. The state of charge estimation for each cycle and route varies for each route and data source. The route includes motorway driving, shows massive energy consumption when the vehicle manages to drive at the highest speed limit and shows less energy consumption when the traffic density restricts the speed. In contrast, the results also show less energy efficiency for city driving when the traffic is dense because the journey time is longer.

The proposed velocity model is valid for generating the potential velocity for the vehicle using any different data source as long as the data is prepared similarly. It can produce real-time velocity profile construction without collecting more data for extended periods, primarily if a more flexible data source is used. It was impossible to integrate weather API and traffic light detection due to the restrictions in the map sources. However, this can be included when using a more flexible API such as OpenStreetMap, even though it has less accuracy compared to others when predicting real-time traffic. Further laboratory experiments will be conducted in the future to validate the results in this paper using the Nissan Leaf battery.

References

1. Conti, J.J., Holtberg, P.D., Beamon, J.A., Michael Schaal, A., Ayoub, J.C., Turnure, J.T.: Annual energy outlook 2014. US Energy Information Administration, p. 2 (2014)
2. Menyah, K., Wolde-Rufael, Y.: CO_2 emissions, nuclear energy, renewable energy and economic growth in the US. Energy Policy **38**(6), 2911–2915 (2010)
3. Wu, X., Freese, D., Cabrera, A., Kitch, W.A.: Electric vehicles' energy consumption measurement and estimation. Transp. Res. Part D: Transp. Environ. **34**, 52–67 (2015)
4. Juul, N., Meibom, P.: Road transport and power system scenarios for northern Europe in 2030. Appl. Energy **92**, 573–582 (2012)
5. Knowles, M., Scott, H., Baglee, D.: The effect of driving style on electric vehicle performance, economy and perception. Int. J. Electr. Hybrid Veh. **4**(3), 228–247 (2012)
6. Zhu, L., Gonder, J.D.: A driving cycle detection approach using map service API. Transp. Res. Part C: Emerg. Technol. **92**, 349–363 (2018)
7. Google Maps API (developers portal). https://cloud.google.com/maps-platform/. Accessed 03 Sept 2019
8. Here Maps. Here Maps APIs (2019). https://developer.here.com/products/maps/
9. TomTom. Tomtom Maps APIs (2019). https://developer.tomtom.com/products/maps-API/
10. Bi, J., Wang, Y., Sai, Q., Ding, C.: Estimating remaining driving range of battery electric vehicles based on real-world data: a case study of Beijing. China. Energy **169**, 833–843 (2019)
11. Kaza, N.: Urban form and transportation energy consumption. Energy Policy **136**, 111049 (2020)
12. Fotouhi, A., Yusof, R., Rahmani, R., Mekhilef, S., Shateri, N.: A review on the applications of driving data and traffic information for vehicles energy conservation. Renew. Sustain. Energy Rev. **37**, 822–833 (2014)
13. Qi, X., Guoyuan, W., Boriboonsomsin, K., Barth, M.J.: Data-driven decomposition analysis and estimation of link-level electric vehicle energy consumption under real-world traffic conditions. Transp. Res. Part D: Transp. Environ. **64**, 36–52 (2018)
14. De Cauwer, C., Van Mierlo, J., Coosemans, T.: Energy consumption prediction for electric vehicles based on real-world data. Energies **8**(8), 8573–8593 (2015)
15. Wager, G., McHenry, M.P., Whale, J., Bräunl, T.: Testing energy efficiency and driving range of electric vehicles in relation to gear selection. Renew. Energy **62**, 303–312 (2014)

16. Yao, E., Yang, Z., Song, Y., Zuo, T.: Comparison of electric vehicle's energy consumption factors for different road types. Discret. Dyn. Nat. Soc. **213** (2013)
17. Bingham, C., Walsh, C., Carroll, S.: Impact of driving characteristics on electric vehicle energy consumption and range. IET Intel. Transport Syst. **6**(1), 29–35 (2012)
18. Barlow, T.J., Latham, S., McCrae, I.S., Boulter, P.G.: A reference book of driving cycles for use in the measurement of road vehicle emissions (2009)
19. Wang, H., Zhang, X., Ouyang, M.: Energy consumption of electric vehicles based on real-world driving patterns: a case study of Beijing. Appl. Energy **157**, 710–719 (2015)
20. Alateef, S., Thomas, N.: Battery models investigation and evaluation using a power demand generated from driving cycles. In: Proceedings of the 12th EAI International Conference on Performance Evaluation Methodologies and Tools, pp. 189–190 (2019)
21. Silvas, E., Hereijgers, K., Peng, H., Hofman, T., Steinbuch, M.: Synthesis of realistic driving cycles with high accuracy and computational speed, including slope information. IEEE Trans. Veh. Technol. **65**(6), 4118–4128 (2016)
22. Yang, Y., Zhang, Q., Wang, Z., Chen, Z., Cai, X.: Markov chain-based approach of the driving cycle development for electric vehicle application. Energy Procedia **152**, 502–507 (2018)
23. Lipar, P., Strnad, I., Česnik, M., Maher, T.: Development of urban driving cycle with GPS data post processing. Promet-Traffic Transp. **28**(4), 353–364 (2016)
24. Esteves-Booth, A., Muneer, T., Kirby, H., Kubie, J., Hunter, J.: The measurement of vehicular driving cycle within the city of Edinburgh. Transp. Res. Part D: Transp. Environ. **6**(3), 209–220 (2001)
25. Iora, P., Tribioli, L.: Effect of ambient temperature on electric vehicles' energy consumption and range: model definition and sensitivity analysis based on Nissan Leaf data. World Electr. Veh. J. **10**(1), 2 (2019)
26. Nissan Leaf. Ev-database (2019). https://ev-database.uk/car/1656/Nissan-Leaf
27. Rahimi-Eichi, H., Chow, M.-Y.: Big-data framework for electric vehicle range estimation. In: 40th Annual Conference of the IEEE Industrial Electronics Society, IECON 2014, pp. 5628–5634. IEEE (2014)
28. Maia, R., Silva, M., Araújo, R., Nunes, U.: Electrical vehicle modeling: a fuzzy logic model for regenerative braking. Expert Syst. Appl. **42**(22), 8504–8519 (2015)
29. Murnane, M., Ghazel, A.: A closer look at state of charge (SOC) and state of health (SOH) estimation techniques for batteries. Analog Devices **2**, 426–436 (2017)

A Robust Approximation for Multiclass Multiserver Queues with Applications to Microservices Systems

Siyu Zhou and Murray Woodside[✉] (iD)

Carleton University, Ottawa K1S 5B6, Canada
`cmw@sce.carleton.ca`

Abstract. Model-based management of software applications in the cloud is based on predicted delays at scaled out services. These services are modeled as FIFO (first-in first-out) multiservers, with many servers, users and types of operation (classes of service). Efficient approximations for these multiservers either scale badly for large systems, or have convergence and accuracy problems. This work investigates three scalable approximations in depth. The best (called AB) combines class aggregation and a binomial approximation to the queue state (which assumes that users behave independently). Over the parameters of greatest relevance, two-thirds of the errors are less than 5%. The largest errors, up to about 30%, occur near the onset of saturation.

Keywords: Queueing · Multiservers · Microservices · Cloud computing

1 Introduction

Model-based management of services can react quickly to changes and can coordinate the scaling of multiple services, as demonstrated in [12]. However, because a scaled-out service is a multiserver, it depends on efficient and robust solution methods for multiserver queues. Existing methods do not meet all the requirements of (i) accuracy, (ii) robust solution in terms of dependable convergence of iterative steps, and (iii) fast solution, and they have not been fully evaluated. This work fills that gap.

The problem is illustrated by the Sock Shop microservices example used in [12]. Six services have sets of load-balanced replicas. The model in Fig. 1 is a layered queueing network (LQN) [8, 10, 11] adapted from [12] with three sets of Clients. The multiplicity of clients and services is shown in curly brackets such as {225} or (for a variable) {$m}. The small nested parallelograms indicate operations with CPU demands in ms. (e.g. [1.2]), and arrows represent calls. The LQNS solver uses iterative Approximate Mean Value Analysis (AMVA) (e.g., [2]) as an overall strategy with options for different algorithms for the multiservers, having different tradeoffs of accuracy and speed.

K. Gilly and N. Thomas (Eds.): EPEW 2022, LNCS 13659, pp. 54–68, 2023.
https://doi.org/10.1007/978-3-031-25049-1_4

Table 1 shows the LQNS solution time for three multiserver options, "Conway" (preferred for accuracy), "Rolia" or "RF" (preferred for speed) and "AB", which was developed in this research and is new. Conway takes much the longest and scales badly in general; Rolia is fast, and AB is faster. A second difficulty with the Rolia option, which motivated this research, is weak robustness due to non- convergence.

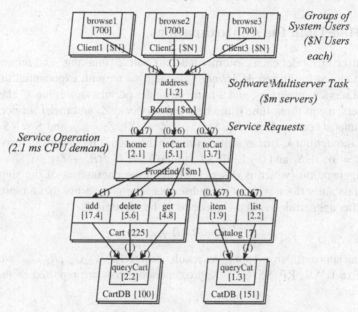

Fig. 1. A layered queueing model of the SockShop Microservices demo [12]

Table 1. Comparison of LQNS solution times of the SockShop model ($\$N = 7000$, $\$m = 110$)

Approximation	Conway	AB	Rolia (RF)
Complexity per iteration (for N_c customers in class c, m servers, C classes)	$O\left(C^3 \prod_c N_c\right)$	$O(m+C)$	$O(C)$
Solution time (seconds)	928.7	0.118	1.426

The goal of this research is to evaluate approximations for multiclass FIFO queues that are heterogeneous (different service times by class), scalable (in the sense of, insensitive to the number of customers) and more robust than RF. They combine scalable single-class approximations which were previously described in [23] with class aggregation. Three approximations are considered here:

- AB, based on a binomial approximation to the queue state distribution and described for a single customer class in [22, 23]
- SS, an equivalent single server (not, at this time, implemented in LQNS and so not shown in Table 1).

- RF, the Rolia algorithm as modified by Franks [8].

The effectiveness of class aggregation in this context, and the accuracy of AB, SS and RF when used with aggregation, were evaluated by comprehensive experiments covering the model parameter space. AB is both the fastest and the most accurate, and is robust. RF is the least accurate and not robust.

2   The Model and the Approximations

In the multiserver model each customer has two states, "thinking" and requesting service. It is a closed multiclass $M/M/m/./N$ queue (that is, with exponential think time, exponential service, m servers, and a finite customer population). It has C classes each with N_c users, mean think time (time between services) Z_c and mean service time S_c, giving parameter vectors $N = (N_1, N_2, \dots)$, $Z = (Z_1, Z_2, \dots)$, and $S = (S_1, S_2, \dots)$. Class c has throughput λ_c, mean waiting time (in the queue) W_c and mean response time R_c, with $R_c = W_c + S_c$ and (by Little's formula) $\lambda_c = N_c/(R_c + Z_c)$.

After aggregation (which is described below) the parameters of the single representative class are written without the subscripts. The probability that a representative customer after aggregation is in the queue or at the server is P:

$$P = R/(R + Z).$$

Applying approximation Appr gives results denoted as $W_{c,\text{Appr}}, R_{c,\text{Appr}}$, where Appr is one of Exact, AB, RF, SS, Sim. Approximation errors are reported as the relative error RE. For class c:

$$RE_c(\text{Appr}) = \left(W_{c,\text{Appr}} - W_{c,\text{Exact}}\right)/R_{c,\text{Exact}}$$

Without a subscript, RE refers to the representative single class. The notation ARE is used for the absolute RE, and $MARE$ for the mean absolute RE over a set of cases.

In designing experiments the traffic intensity is set by choosing values of a load intensity value T which is defined as the ratio of the maximum arrival rate N/Z to the maximum service rate m/S:

$$T = NS/(mZ).$$

The response time function has a "knee" near to $T = 1$, where it turns up as traffic increases, and this is where the largest errors were found. Figure 2 illustrates the close relationship between T and the server saturation defined as saturation = utilization/m. The upper curves are for smaller values of m and N, and the lower curves for larger values. For large N, the two have nearly equal values from zero up to near unity, and T then increases without limit as saturation approaches 1.0.

Since service autoscaling typically keeps the saturation below about 0.9, values of T between 0 and 2 seem to be of the greatest practical interest. Higher values may occur at heavily saturated bottlenecks without autoscaling.

For multiclass evaluation experiments the per-class intensities were set to a value $T'_c = T_{nominal}/C$, where $T_{nominal}$ is chosen as a nominal total intensity. The resulting intensity T for the representative class was then found to be close to $T_{nominal}$.

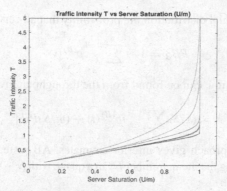

Fig. 2. The Relationship of the Traffic Intensity to Server Saturation for an M/M/m Queue, for all combinations of $m = [2, 5, 20]$ and $N = [10, 100, 1000]$

2.1 The Approximations for One Class

For the representative class with parameters N, S, and Z, the approximations are:

The Rolia-Franks Approximation (RF, Giving Approximation W_{RF}): ([8, 15] Ch. 6) assumes independent servers to estimate the probability PB that all servers are busy. Using this PB, it estimates waiting at each iteration by:

$$W = S + PB(S/m)L^*, \tag{1}$$

where L^* is the expected customers when N is replaced by $N - 1$, as in conventional AMVA [2]. In fixed-point iteration, convergence requires under-relaxation of the form

$$\text{updated } W = \alpha \text{ (new } W \text{ as in Eq.(1))} + (1 - \alpha) \text{ (previous } W)$$

with a relaxation parameter α less than unity. The complexity of RF combined with AMVA by Proportional Estimation (PE) [18] or by Linearizer [5] is O(1) per iteration for each multiserver queue, making RF scalable in our sense.

The Equivalent Single Server (SS, Giving W_{SS}): Some reports on model-based auto-scaling (e.g. [20, 21] have approximated a set of m servers by a faster single server with service time S/m. SS also adds an additional delay $S(1 - 1/m)$ to provide the correct total delay at light loads, which is not found in the references. The additional delay is added to the response time of the server, but does not contribute to the server utilization. The model is solved as a single server by AMVA. Using PE the time complexity of each iteration is O(1), and with Linearizer [5] It is larger but also O(1).

The Arrival-Theorem Binomial Approximation (AB, giving W_{AB}): AB assumes that the movement of customers between the thinking and server states is independent, giving a binomial distribution with probability $p^{(B)}(i)$ of i customers at the queue and server. This is the single-class version of the approach by deSouza e Silva and Muntz described in [19]. It gives:

$$p^{(B)}(i) = [N!/(N - i)!i!]P^i(1 - P)^{N-i}.$$

Based on the binomial distribution the probability that all servers are busy is PB_B:

$$PB_B = 1 - \sum_1^{m-1} p^{(B)}(i).$$

The mean waiting time can be found from the throughput, given by:

$$\lambda = (1/S)\sum_1^{m-1} ip^{(B)}(i) + (m/S)PB_B,$$

however this direct approach gives inferior estimates. AB instead applies the Arrival Theorem [14] by finding the mean number L^* in a queue with $N-1$ customers, indicated by a superscript $(N-1)$:

$$L^* = \sum_{i=m}^{N-1} (i - m + 1)p_{AB}^{(N-1)}(i).$$

This can be re-arranged so that it uses only the first m probabilities:

$$L^* = (N - 1)P - (m - 1)\left(1 - \sum_{i=1}^{m-1} p^{(N-1)}(i)\right) - \sum_{i=1}^{m-1} ip^{(N-1)}(i). \quad (2)$$

Then by the Arrival Theorem, a customer waits on average for L^* departures, giving

$$W_{AB} = L^* S/m. \quad (3)$$

Using Eq. (2) the complexity of AB is $O(m)$. Since it does not depend on N it is also scalable in our sense.

These three approximations were compared for a single class of customers in [23], from which Fig. 3 illustrates the relationship of the errors to the traffic intensity T for cases with $N = 100$, $S = 1$, $m = 3$, 10 and 30 and Z set to $N/(mT)$. The relative error RE is largest around $T = 1.0$ and approaches zero for light and very heavy traffic.

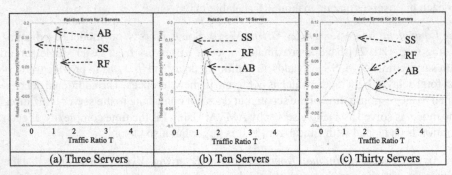

| (a) Three Servers | (b) Ten Servers | (c) Thirty Servers |

Fig. 3. Examples of the Relative Errors of the AB, SS and RF Approximations (from [23])

3 Class Aggregation

Class aggregation is supported by the observation that the waiting times of different classes at a FIFO multiserver are usually not very different [13]. Especially if the populations of the classes are all quite large, it is intuitively reasonable that an arrival of any class will see a closely similar mix of classes in the queue, and the FIFO discipline treats all classes equally. This is made more precise in the following hypothesis.

Interclass Waiting Difference Hypothesis: The maximum difference between the mean class waiting times at a FIFO queue is the maximum of the class service times at the queue:

$$\max_{c,d}(|W_c - W_d|) \le \max_c(S_c).\tag{4}$$

A heuristic argument for this hypothesis assumes (1) the Arrival Theorem and (2) that the time to the first departure is the same for all classes. Let Q represent the mean total work in the queue (excluding the server) at steady state, and Q_c, the same quantity if one customer in class c is removed. By the Arrival Theorem the difference between the waiting times of two classes c and d is the difference between Q_c and Q_d:

$$W_c - W_d = Q_c - Q_d.$$

Q_c is less than Q by an amount due to the removal of the customer, an amount that may be reduced by contributions from other classes. This suggests that

$$Q - S_c \le Q_c \le Q \text{ and } Q - S_d \le Q_d \le Q.$$

The hypothesis follows at once by considering the rectangular set of points (W_c, W_d) defined by the inequalities.

The hypothesis was also verified experimentally on 500 cases spanning the parameter space, using the Conway solution for cases where it gave a value in reasonable time and simulation otherwise, and it was satisfied for every case.

Aggregation Solution. The aggregation gave a single representative class with parameters weighted by the relative throughputs, as:

$$N = \sum_c N_c; \; \lambda = \sum_c \lambda_c; \; S = \sum_c \lambda_c S_c/\lambda; \; Z = \sum_c \lambda_c Z_c/\lambda; \; T = NS/(mZ)$$

The class throughput depends on the approximate waiting W (using Little's formula), as:

$$\lambda_c = N_c/(W + S_c + Z_c).\tag{5}$$

Because the throughput depends on W the calculation is iterative, with the following outline algorithm:

High-level Multiclass Waiting-time Algorithm
```
input vectors N, S, Z.
```
initialize $N = \sum_c N_c$, $W = \sum_c N_c S_c / 2$, $eps = 10^{-6}$, $relax = 0.7$
```
repeat until difference < eps
       set λc = Nc/(W + Sc + Zc) for each class c (Eq. (2))
       find the representative values N, S and Z from Eq
   (EQ)
       find Wnew by the approximation AB, SS, or RF
```
$\qquad difference = |W_{new} - W|$
$\qquad W = relax * W_{new} + (1 - relax) * W$
```
return W
```

3.1 Evaluation Cases

As there is no exact solution, the relative errors were found by simulation (with 95% confidence intervals less than 1% of the estimated waiting). Experiments were performed for two, four and eight classes, with 720 cases each. To create each set, all combinations of the following values were employed:

- Ten values of vectors **S** and **Z** with components uniformly distributed over [0, 1],
- Three values of m in the set [2, 4, 8],
- 24 values of the nominal traffic intensity $T_{nominal}$ with values concentrated in the interval (0,2) of greatest interest: [0.02, 0.1, 0.2, 0.3, 0.4, 0.5, 0.6, 0.7, 0.8, 0.84, 0.9, 0.96, 1, 1.1, 1.2, 1.3, 1.4, 1.5, 1.6, 2, 3, 4, 6]

For each combination, N_c for each class was set according to the traffic intensity, as

$$N_c = \max(1, \lceil (T_{nominal}/C) Z_c m / S_c \rceil).$$

4 The Approximation Error

Table 2 summarizes the absolute relative errors, including the single-class results reported in [23]. AB has the smallest ARE, and RF the largest. RF also had many convergence failures (fewer if the relaxation factor was reduced, but this also slows down the solver). The maximum errors were close to 50% for AB and SS and nearly 100% for RF (these were due to convergence failures). It is notable that SS is nearly as good as AB on average and has smaller maximum errors.

Table 2. Approximation errors with different numbers of classes

	Measure	Approximation error			RF non-converged
		AB	SS	RF	
One class, 0 < T < 5 (30000 cases, [23])	Mean ARE	0.0089	0.0142	0.0191	14130/30000 (*relax* = 0.7)
	Max ARE	*0.243*	*0.190*	*0.249*	
One class, 5 < T < 36 (30000 cases, [23])	Mean ARE	0.00041	0.00079	0.00307	24030/30000 (*relax* = 0.7)
	Max ARE	*0.0472*	*0.1859*	*0.08792*	
Two classes (720 cases)	Mean ARE	0.04059	0.04132	0.06166	2/720 (*relax* = 0.1)
	Max ARE	*0.4563*	*0.4005*	*0.6249*	
Four classes (720 cases)	Mean ARE	0.03397	0.03895	0.07989	5/720 (*relax* = 0.1)
	Max ARE	*0.5321*	*0.4649*	*0.8061*	
Eight classes (720 cases)	Mean ARE	0.03236	0.03752	0.1191	25/720 (*relax* = 0.1)
	Max ARE	*0.4669*	*0.3962*	*0.9288*	

As classes increase, the mean errors decrease for AB and SS, but increase for RF. The relatively small mean errors reported for a single class reflect the distribution of single-class cases which emphasized large values of T (which gave small errors).

The relationship of the relative error to the traffic intensity is displayed in the scatter plots of Fig. 4. The plots are restricted to $T < 3$; all errors were small for larger T. The error patterns are all similar, and similar to the examples in Fig. 3. Away from T = 1 the errors approach zero fastest for AB and slowest, for RF. Above 30% error there are only nine points for AB, and six points for SS, but many more (with much larger values) for RF. The largest errors for RF were due to non-convergent solutions, which were included in the plots.

The scatter plots tend to conceal the preponderance of points with small errors, which is better seen in the percentile values in Table 3. For AB almost two-thirds of the cases had less than 5% error.

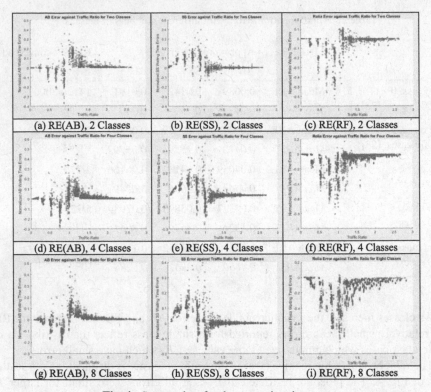

Fig. 4. Scatter plots for the approximation errors

Table 3. Percentiles of absolute relative errors

Range of ARE	AB	SS	RF
0–0.01	0.42	0.38	0.40
0–0.05	0.65	0.57	0.58
0–0.20	0.96	0.98	0.82

4.1 Errors with $T < 2$ and Large N

It was argued in Sect. 2 that practical cases will tend to have $T < 2$ and large N, and another set of cases was examined in this range.

Simulations were run with all combinations of the parameters $N = [100, 200, 300, 400, 500]$, $m = [2, 5, 10, 20, 50]$. T took 10 values between 0.2 and 2.0, and S was random with a mean of 1.0. Figure 5 shows error histograms and Table 4 shows the mean errors for 2, 4 and 8 classes.

Figure 5 shows that when we focus on T less than 2.0, the errors for AB increase with more classes, unlike the results for the wider range of traffic, and the other approximations do not show a clear trend.

AB has errors less than 5% in more than two-thirds of the cases, and less than 20% in 94% or more of the cases. This is somewhat better than SS and very much better than RF. If we consider only 10 or more servers *RE* is much smaller, less than 5% in over 85% of cases, and less than 20% in over 98% of cases.

For less than 10 servers the error increases with the number of classes, possibly due to class aggregation, but with 10 or more servers this was not observed.

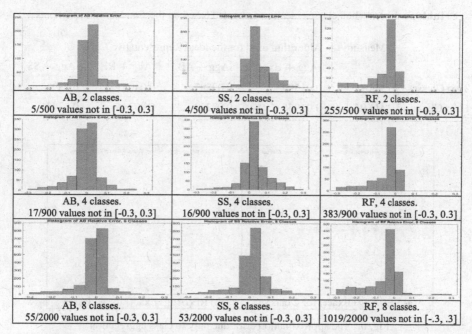

AB, 2 classes.	SS, 2 classes.	RF, 2 classes.
5/500 values not in [-0.3, 0.3]	4/500 values not in [-0.3, 0.3]	255/500 values not in [-0.3, 0.3]
AB, 4 classes.	SS, 4 classes.	RF, 4 classes.
17/900 values not in [-0.3, 0.3]	16/900 values not in [-0.3, 0.3]	383/900 values not in [-0.3, 0.3]
AB, 8 classes.	SS, 8 classes.	RF, 8 classes.
55/2000 values not in [-0.3, 0.3]	53/2000 values not in [-0.3, 0.3]	1019/2000 values not in [-.3, .3]

Fig. 5. Histograms of *RE* for the waiting times of all classes

Table 4. Mean absolute relative errors with populations in [100, 500] and T in [0.2, 2]

	MARE for AB	*MARE* for SS	*MARE* for RF
2 Classes (250 cases)	0.0364	0.0483	0.4246
4 Classes (225 cases)	0.0453	0.0584	0.3569
8 Classes (250 cases)	0.0625	0.0568	0.6329

4.2 Error Component Due to Class Aggregation

To isolate the error due to class aggregation, an "ideally aggregated" model "iAgg" was created using simulation throughputs as weights. *RE*(iAgg + Exact) is the error of aggregation only, using the exact single class solution, while *RE*(iAgg + Appr) is the error of approximation Appr applied to the iAgg model. Table 5 shows that the errors for the two-class and four-class cases are roughly evenly divided between aggregation error and the approximation errors (columns 4, 5, 6 are about double column 3). Scatter plots in Fig. 6 show that the aggregation error alone is largest around $T = 1$, and takes both positive and negative values. The results in column 3 suggest that the aggregation error may increase with C, but the approximations appear to counter that increase, with smaller errors for four classes.

Table 5. Errors when classes ideally aggregated based on the simulation throughputs

	Measure	Algorithm used for the Ideal Representative Class			
		iAgg + Exact	iAgg + AB	iAgg + RF	iAgg + SS
Two Classes	*MARE*	0.01599	0.03887	0.04031	0.03922
Four Classes	*MARE*	0.01966	0.03250	0.03245	0.03431

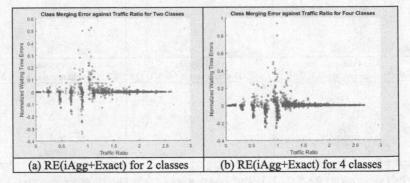

(a) RE(iAgg+Exact) for 2 classes   (b) RE(iAgg+Exact) for 4 classes

Fig. 6. The approximation error due only to class aggregation

5   Application to the SockShop Microservices Model

SockShop is a small but realistic microservices demonstration system. Figure 1 shows a LQN performance model adapted from [12], with three groups of users instead of one, and variable numbers of service replicas. The three approximations implemented in the LQNS solver (AB, RF, Conway) were compared in predicting the performance effect of different scalings of the Router and FrontEnd services, given by the parameter $m in Fig. 1. The value shown of 10 replicas was increased in 15 steps up to 240, reducing the response time as shown in Fig. 7. The performance improves by several orders of magnitude. The AB results were identical to Conway within 3 figures, while Rolia diverged by up to 10% at the lowest value of $m.

The mean CPU time for the solution across the cases was 0.37 s. for AB, and 1.05 s for Rolia. The time for Conway increased steeply with $m up to 28859 s (more than 8 h) at $m = 190, where the experiment stopped. Thus in this case Conway is impractical, while Rolia is less accurate than AB and takes nearly three times as long (although both are quite fast).

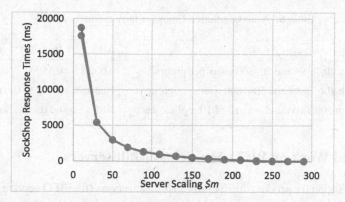

Fig. 7. Response time of the SockShop model with scaled front-end services

6 Scalability of the Approximations to Very Large Models

To consider large models it is not convenient or necessary to use models of real systems. Every possible pattern of components and interactions might be of interest, so models with random structure and parameters were used, with 20, 100 and 200 tasks, generated by the utility program lqngen [9]. Each task offers an average of 10 operations and is scaled to an average of 20 replicas; the model represents a highly complex system. The 100-task case is illustrated in Fig. 8. All cases had 5 groups of users.

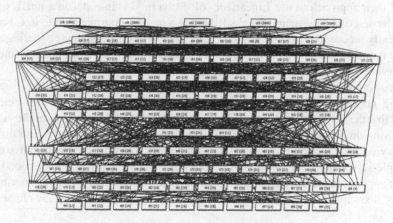

Fig. 8. A random model with 100 tasks.

The models were solved by AB and RF, with Conway for further comparison. SS was not applied because it is not implemented in the solver. The solution times for these models are shown in Table 6.

The AB algorithm is clearly more scalable and more robust than RF. The 200-task model is approaching the storage limits of the solver, and exceeds those of the simulator. The increase in the times for AB is greater than is accounted for by the complexity of AB per server and per iteration, presumably dominated by other operations in LQNS.

Table 6. Average solution times for large random models

Cases	AB	RF	Conway
15 cases with 20 Tasks and 20–300 users per group	0.062 s	0.189 s	14.138 s
20 cases with 100 Tasks and 50–1000 users per group	10.3 s	358.2 s	no result
One case with 200 Tasks and a mean of 10 replicas each	80.5 s	no result	no result

7   Related Work on FIFO Multiclass Multiservers

The exact solution for product-form multiclass multiservers (for FIFO service with equal class service times or for processor-sharing queueing with equal or unequal times) is described in [14]. For FIFO service and unequal times, AMVA approximations are described in [1, 5] and (by recasting the solution as an optimization problem) in [3].

Recently Legato and Mazza [13] have solved these networks using class aggregation, giving a single class queueing network which they solve by exact MVA. They go further in aggregation than our SS approximation, which keeps the classes separate except for the calculation of waiting times at multiservers. However, all of these methods apply to FIFO servers only if the mean class service times are equal.

For unequal class service times, Ruth [16] adapted exact MVA; Schmidt [17] approximated the multiserver by a (product-form) state-dependent-rate server; de Souza e Silva and Muntz approximated the queue state by a multinomial distribution [19]; and Conway adapted their approach to use Linearizer [6]. Rolia in [15] introduced a much simpler AMVA as described above and Franks [8] made a modest improvement to it. Casale [4] has recently proposed replacing any server by an approximate processor-sharing server updated iteratively based on properties of a diffusion model; the work does not mention multiservers but appears to be capable of generalization in that direction.

Regarding class aggregation, Dowdy et al. [7] considered aggregated classes in queueing networks in which class service times at FIFO servers are the same, and showed that the aggregate analysis understates the total throughput. This result suggests that waiting time errors due to aggregation should be positive, however with different class service times the results reported in Sect. 4 show both positive and negative errors.

Single-server approximations similar to SS have been used (without the added delay term) in applied studies such as [20, 21], but only for a single class. Their approximation accuracy does not appear to have been studied previously, for one or many classes.

8   Conclusions

All the approximations which are scalable in the sense that their time complexity does not increase with the customer population have similar patterns of error. The results in Sect. 4.1 indicate that errors are substantial only in an interval of the traffic intensity T between 0.5 and 1.5, and they do not exceed a relative value of about 30%. All the approximations had asymptotic errors under light and heavy traffic which rapidly approached zero. For AB and RF the approach as T diverged from unity was more rapid than for SS.

The practical usability of model-based autoscaling using these approximations may be limited by these errors, since the stated traffic levels include the intended operating point of most autoscaling systems. Model-based autoscaling would have to tolerate this range of prediction errors in the response time of individual services. Fortunately in a large system with many scaled services the errors will tend to average out, so the overall QoS control may be considerably better than this. Also predictions regarding more heavily loaded services, which are the most important, are more accurate.

Practical usability also implies a preference for AB or SS over RF, due to non-convergence of RF in some situations, particularly for large T.

The errors for heterogeneous multiclass servers are contributed partly by class aggregation and partly by the use of the approximation for the single representative (i.e. aggregated) class. The results in Sect. 4.2 indicate that the two contributions are roughly equal.

The impact of the number of classes C on the error depends on the traffic levels. For T in a broad range (in Table 2) the average error decreases with C for AB and SS, and increases for RF. For T near to unity (in Table 4) it increases for AB, and shows no clear trend for SS and RF. The impact of the number of servers, m, is also complex. However if m is larger than about 10, then the error decreases with further increases in m, for all approximations.

An unexpected result is that SS is quite close to AB in accuracy. Clearly it is much the simplest to calculate, so it might be the algorithm of choice. The additional delay term in SS has to be implemented with care since it is part of the response time of the server.

References

1. Akyildiz, I.F., Bolch, G.: Mean value analysis approximation for multiple server queueing networks. Perform. Eval. **8**, 77–91 (1988)
2. Bolch, G., Greiner, S., Meer, H., Trivedi, K.: Queueing Networks and Markov Chains: Modeling and Performance Evaluation with Computer Science Applications, 2nd edn. Wiley, Hoboken (2006)
3. Casale, G., Perez, J., Wang, W.: QD-AMVA: evaluating systems with queue-dependent service requirements. Perform. Eval. **91**, 80–98 (2015)
4. Casale, G.: Integrated performance evaluation of extended queueing network models with line. In: Proceedings of the Winter Simulation Conference, pp. 2377–2388 (2020)
5. Chandy, K.M., Neuse, D.: Linearizer: a heuristic algorithm for queueing network models of computing systems. Comm. of the ACM **25**(2), 126–134 (1982)
6. Conway, A.E.: Fast approximate solution of queueing networks with multi-server chain-dependent FCFS queues. In: Modeling Techniques and Tools for Computer Performance Evaluation, Plenum, New York, pp 385–396 (1989). https://doi.org/10.1007/978-1-4613-0533-0_25
7. Dowdy, L.W., Carlson, B.M., Krantz, A.T., Tripathi, S.K.: Single-class bounds of multi-class queuing networks. J. ACM **39**(1), 188–213 (1992)
8. Franks, G.: Performance analysis of distributed server systems. Ph.D thesis, Carleton University (1999)
9. G. Franks, G.: lqngen − generate layered queueing network models. https://github.com/layeredqueuing/V5. Accessed 10 Feb 2022

10. Franks, G., Al-Omari, T., Woodside, M., Das, O., Derisavi, S.: Enhanced modeling and solution of layered queueing networks. IEEE Trans. Software Engineering **35**(2), 148–161 (2009)
11. Franks, G. et al: Layered Queueing Network Solver and Simulator User Manual, Carleton University. http://www.sce.carleton.ca/rads/lqns/userman22.pdf. Accessed 20 Jan 2022
12. Gias, A.U., Casale, G., Woodside, M.: ATOM: model-driven autoscaling for microservices. In: 39th International Conference on Distributed Computing Systems (ICDCS), pp 1994–2004 (2019)
13. Legato, P., Mazza, R.M.: Class aggregation for multi-class queueing networks with FCFS multi-server stations. In: Phung-Duc, T., Kasahara, S., Wittevrongel, S. (eds.) QTNA 2019. LNCS, vol. 11688, pp. 221–239. Springer, Cham (2019). https://doi.org/10.1007/978-3-030-27181-7_14
14. Reiser, M., Lavenberg, S.S.: Mean-value analysis of closed multichain queueing networks. J. ACM **27**(2), 312–322 (1980)
15. Rolia, J.A., Sevcik, K.C.: The method of layers. IEEE Trans. Softw. Eng. **21**, 689–700 (2015)
16. Ruth, A.: Entwicklung, Implementierung und Validierung neuer Approximationsverfahren fur die Mittelwertanalyse (MWA) zur Leistungsberechnung von Rechnersystemen. Diplomarbeit am IMMD der Friedrich-Alexander-Universitat Erlangen-Nurnberg (1987)
17. Schmidt, R.: An approximate MVA algorithm for exponential, class-dependent multiple servers. Perform. Eval. **29**, 245–254 (1997)
18. Schweitzer, P.J.: Approximate analysis of multiclass closed networks of queues. In: Proceedings of the International Conference on Stochastic Control and Optimization, Amsterdam, pp 25–29 (1979)
19. Silva, E.D.S., Muntz, R.R.: Approximate solutions for a class of non-product form queueing network models. Perform. Eval. **7**, 221–242 (1987)
20. Zhang, Q., Xiao, Y., Liu, F., Lui, J.C.S., Guo, J., Wang, T.: Joint optimization of chain placement and request scheduling for network function virtualization. In: International Conference on Distributed Computing Systems (ICDCS), pp 731–741 (2011)
21. Zhang, Q., Zhu, Q., Zhani, M.F., Boutaba, R., Hellerstein, J.L.: Dynamic service placement in geographically distributed clouds. IEEE J. Sel. Areas Commun. **31**, 762–772 (2013)
22. Zhou, S.: A New Approximation for Multiserver Waiting Time for Layered Queueing Systems. MASc thesis, Carleton University (2021)
23. Zhou, S., Woodside, M.: A multiserver approximation for cloud scaling analysis. In: Proceedings of the Workshop on Challenges in Software Performance (WOSPC-22), in the Companion Volume to the International Conference on Performance Engineering (ICPE22), ACM, New York (2022)

Stochastic Modelling

Product Form Solution
for the Steady-State Distribution
of a Markov Chain Associated
with a General Matching Model
with Self-loops

Ana Busic[1], Arnaud Cadas[1], Josu Doncel[2], and Jean-Michel Fourneau[3(⊠)]

[1] INRIA and DIENS, École normale supérieure, PSL University, CNRS, Paris, France
ana.busic@inria.fr
[2] Univ. of the Basque Country, UPV/EHU, Leioa, Spain
josu.doncel@ehu.eus
[3] DAVID, Univ. Paris-Saclay, UVSQ, Versailles, France
Jean-Michel.Fourneau@uvsq.fr

Abstract. We extend the general matching graph model to deal with matching graph where every node has a self loop. Thus the states on the Markov chain are associated with the independent sets of the matching graph. We prove that under i.i.d. arrivals assumptions the steady-state distribution of the Markov chain has a product form solution.

1 Introduction

Intuitively, a Matching model describes the waiting times suffered by items before they match and disappear immediately once they are matched. It is an easy representation of multiple types Rendez-Vous between items. Following [1] a Matching model is a triple (G, Φ, μ) formed by

1. a matching graph $G = (\mathcal{V}, \mathcal{E})$ which is an undirected graph whose vertices in $\mathcal{V}$ are classes of items and whose edges in $\mathcal{E}$ models the allowed matching of items. G is called the compatibility graph or the Matching graph.
2. Φ is a matching policy. It states the couple of items which is chosen upon arrival when an arriving item matches one or more types of item already waiting.
3. a distribution of probability μ to model the arrivals of items. Alternatively one can consider a collection of Poisson processes for a continuous-time model.

The Matching graph represents the classes of items and the compatibility among classes of items. Upon arrival, an item is queued if there are not compatible items present in the system. A matching occurs when two (or more) compatible items are present and it is performed according to the matching discipline. Typical matching disciplines are First Come First Match (an analog of First Come First

K. Gilly and N. Thomas (Eds.): EPEW 2022, LNCS 13659, pp. 71–85, 2023.
https://doi.org/10.1007/978-3-031-25049-1_5

Served in this approach) or Match the Longest Queue. Once they are matched, both items leave the system immediately (no need for service). Note that even if a Matching model has a queueing theory flavor, the items play the roles of both customers and servers. In some sense it has also some links with two well known stochastic models: networks with positive and negative customers and stochastic Petri nets. In a network with positive and negative customer proposed by Gelenbe in [2] a negative customer can provoke the instantaneous deletion of a positive customer but it is never queued in the network. In [3] the deletion depends on the type as in a matching while in [4] customers do not receive service but wait until they are deleted by negative customers. In a stochastic Petri net, tokens wait until they match but the places where the tokens match are usually associated with non negative delays [5]. Under some structural conditions, stochastic Petri nets may have a product form steady-state solution (see for instance [6,7]). Even if it possible to associate a Petri net with a matching graph model, our results differ from [6,7].

Despite its simple formulation, Matching models were not so simple to analyze. Assuming independent Poisson arrivals of items, and FCFM discipline the model is associated with an infinite Markov chain. Under these assumptions, a necessary condition of stability and a product form solution were proved in [8] and [1]. Moreover we recently established that there exists some performance paradox for FCFM matching models [9]. When one add new edges in the compatibility graph, one may expect that the expectation of the total number of customers decreases. In [9] we have given some examples which show that it is not always the case and we prove a sufficient condition for such a performance paradox to exist. Thus adding flexibility on the matching does not always result in a performance improvement.

The general matching model proposed in [8] and [1] was considering a general undirected matching graph G and it is assumed that the arrivals of items occur one at a time. It is important to avoid the confusion with Bipartite Matching Model (see for instance [10] and references therein) where the matching graph is bipartite and two items of distinct classes arrive at the same time. Bipartite Matching Models were motivated by analysis of the public housing [11]. In this model, households which apply for public housing and housings which become available both arrive over time. Once the matching is done, the housing is occupied for a long time period. Thus it is more convenient to represent them as an arrival streams of items rather than traditional servers in a queue. Another application studied in the literature was the kidney exchanges [12,13]. The kidney exchange arises when a healthy person who wishes to donate a kidney is not compatible (blood types or tissue types) with the receiver. Two incompatible pairs (or maybe more) can form a cyclic exchange, so that each patient can receive a kidney from a compatible donor (see [14] and reference therein for a presentation of the problem and the modeling and algorithmic issues).

Here, we further extend the type of graph to represent the general matching. We assume that all the nodes in the matching graph have a self loop (see Fig. 1 for an example of such a graph) and this was clearly forbidden by the previous assumptions in [1]. Note that the results obtained in [8] and [1] are not valid anymore because of some technical details in the proofs of product form. It is required in these papers that the matching graph does not contain self loops. Without self loops the Markov chain is infinite and the system stability has to be studied taking into account the matching discipline.

For our new model, the stability problem is not an issue. As all the nodes in the compatibility graph have a self loop, the system can contain at most one item of each type. Therefore the Markov chain associated with the population is finite. If the chain is irreducible, it is therefore ergodic (see more details in Sect. 2). Such a model may represent how we can organize fair competition between players with roughly the same ranking (for instance ELO points for chess). Such an application was not possible for the model described in [1].

A more general result for matching with multigraph has recently been presented in [15]. In that paper a multigraph is defined as a graph where the self loops are allowed. Therefore the proof they proposed is more general than our result because the self loops are not mandatory leading to potential infinite Markov chains. However our proof is based on simpler arguments (balance equations rather than reversibility of an extended process followed by an aggregation) and we hope it has its own value.

The technical part of the paper is as follows. We begin in the next section with the notations. Section 3 is devoted to a simple example built for a matching graph with 4 nodes. We obtained the steady-state distribution and checked that a multiplicative solution holds. In Sect. 4, we prove that this product form solution holds for every matching graph with self loops for all the nodes.

2   Notation and Assumptions

Let $G = (\mathcal{V}, \mathcal{F})$ be the matching graph. Nodes in $\mathcal{V}$ are also denoted as letters. An ordered list of letters is called a word. Assume that x is a letter from $\mathcal{V}$, $\Gamma(x)$ is the set of neighbors of x in G. As all nodes in $\mathcal{V}$ carry a loop, we have $x \in \Gamma(x)$ for all node x.

Let m be an arbitrary word.

- $(m|x)$ is the word obtained by appending letter x at the end of word m while $(m + x)$ represents the set of words obtained by adding an x anywhere inside word m (even before word m).
- $|m|$ will denote the size of word m (i.e. the number of letters).
- $m(\psi)$ will be the letter of m at position ψ.
- $\Gamma(m) = \cup_{x \in m} \Gamma(x)$.
- $Pre(m, \psi)$ is the prefix of m with size ψ. The prefix of size 0 is the empty word denoted as E or $\emptyset$.
- Similarly $Suf(m, \psi)$ is the suffix of m with size ψ.
- Finally, $In(m, x, \psi)$ is the word of size $|m| + 1$ built from m after the insertion of x at position ψ.

We consider a discrete time model. We assume i.i.d. arrivals of a letter at every time slot. We consider the First Come First Match policy or FCFM (sometimes denoted as the First Come First Served discipline in the literature). An arriving letter will be added at the end of the word if it does not match any letter in the current word. If the arriving letter matches one or several letters of the current word, both the oldest matching letter and the arriving letter vanish immediately.

α_i will denote the probability of arrival of letter i while α_0 is the probability that there is no arrival. The state of the system at time t is a word. Under these assumptions, the process $\{m_t\}$, with $t \in \mathbb{N}$ is a Discrete Time Markov chain. If $\alpha_0 > 0$, this Markov chain is aperiodic. Furthermore it is clear that all the states of the Markov chain are independent sets of the matching graph. Indeed, if two letters are neighbors in the matching graph, they cannot be in a state of the Markov chain. Furthermore, as all the nodes of the matching graph have a self loop, it is not possible to have several occurrences of the same letter in a node of the chain. This last property does not hold for matching models without loops (see for instance [8]). Therefore the Markov chain is finite (again this is not true in the model studied in [8] and [1]). Clearly if $\alpha_i > 0$ for all i, the chain is irreducible. Therefore we do not have to study the stability problem as in [1]. The chain is always ergodic.

For all subset S of nodes of G, α_S will denote the probability of arrival of letters in S:

$$\alpha_S = \sum_{x \in S} \alpha_x$$

To simplify the formulation of the steady-state distribution, we make the following remark:

Remark 1. *Let m be an arbitrary state and x be an arbitrary letter, such that $(m|x)$ is a valid state of the chain, we have $\Gamma(m + x) = \Gamma(m|x) = \Gamma(x|m)$. Remember that $\Gamma(x)$ is the set of neighbors of state x.*

3   Path of Length 4

We begin with a simple example. We consider as a matching graph, a path of length 4 (usually denoted as $P4$) with loops on every node (see Fig. 1).

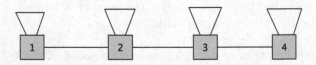

Fig. 1. Matching graph: $P4$.

We first build the Markov chain associated with this matching graph. The states are based on the independent sets of this graph. The states take into account the order of arrivals (remember that we consider FCFM discipline). These states (and sets) contain up to two letters. A state is labelled by the letters which are included while E will denote the empty state. $(x|y)$ represents the state containing letter x followed by letter y. The associated independent set is $\{x, y\}$. Therefore set $\{x, y\}$ is associated with states $(x|y)$ and $(y|x)$.

This Markov chain has 11 states: E, 1, 2, 3, 4, $(1|4)$, $(4|1)$, $(1|3)$, $(3|1)$, $(2|4)$, $(4|2)$. The graph of the Markov chain obtained by the XBorne [16] tool is depicted in Fig. 2. We do not add the transition probabilities but the arrival labels to make the figure more understandable. For the same reason, we do not draw either the loop on every state associated with the null arrival event (with probability α_0).

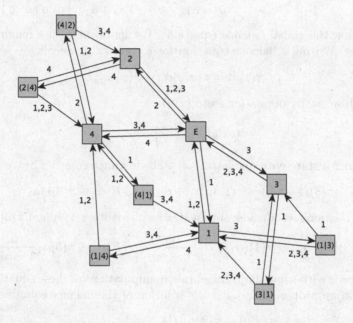

Fig. 2. Graph of the Markov chain associated with Matching graph $P4$.

We will write the global balance equations for an arbitrary node of the Markov chain, taking into account the graph properties of the nodes. We consider the following partition of the states of the chain based on their properties.

1. State E.
2. States associated with maximal independent sets. For this matching graph, a maximal independent set contains two letters and they are denoted $(x|y)$.
3. States which are neither the empty state nor a state associated with a maximal independent set. Here, these states are words with a single letter.

A similar partition will be used in the next section to make the proof in the general case.

To write the transition probability matrix, we consider that the states are in the following order E, 1, 2, 3, 4, $(1|4)$, $(4|1)$, $(1|3)$, $(3|1)$, $(2|4)$, $(4|2)$. The partition is added into the matrix presentation to make the block structure more visible.

$$P = \begin{bmatrix} \alpha_0 & \alpha_1 & \alpha_2 & \alpha_3 & \alpha_4 & 0 & 0 & 0 & 0 & 0 & 0 \\ \alpha_1+\alpha_2 & \alpha_0 & 0 & 0 & 0 & \alpha_4 & 0 & \alpha_3 & 0 & 0 & 0 \\ \alpha_1+\alpha_2+\alpha_3 & 0 & \alpha_0 & 0 & 0 & 0 & 0 & 0 & 0 & \alpha_4 & 0 \\ \alpha_2+\alpha_3+\alpha_4 & 0 & 0 & \alpha_0 & 0 & 0 & 0 & 0 & \alpha_1 & 0 & 0 \\ \alpha_3+\alpha_4 & 0 & 0 & 0 & \alpha_0 & 0 & \alpha_1 & 0 & 0 & 0 & \alpha_2 \\ 0 & \alpha_3+\alpha_4 & 0 & 0 & \alpha_1+\alpha_2 & \alpha_0 & 0 & 0 & 0 & 0 & 0 \\ 0 & \alpha_3+\alpha_4 & 0 & 0 & \alpha_1+\alpha_2 & 0 & \alpha_0 & 0 & 0 & 0 & 0 \\ 0 & \alpha_3+\alpha_4 & 0 & \alpha_1+\alpha_2 & 0 & 0 & 0 & \alpha_0 & 0 & 0 & 0 \\ 0 & \alpha_2+\alpha_3+\alpha_4 & 0 & \alpha_1 & 0 & 0 & 0 & 0 & \alpha_0 & 0 & 0 \\ 0 & 0 & \alpha_4 & 0 & \alpha_1+\alpha_2+\alpha_3 & 0 & 0 & 0 & 0 & \alpha_0 & 0 \\ 0 & 0 & \alpha_3+\alpha_4 & 0 & \alpha_1+\alpha_2 & 0 & 0 & 0 & 0 & 0 & \alpha_0 \end{bmatrix}$$

Consider now the global balance equations. Let us begin with a maximal independent set. Writing a balance equation for a state $(x|y)$ we get:

$$\pi(x|y) = \pi(x|y)\alpha_0 + \pi(x)\alpha_y,$$

from which we easily obtain for state $(x|y)$:

$$\pi(x|y) = \pi(x)\frac{\alpha_y}{1-\alpha_0}.$$

Now consider a state with one letter (x). For instance consider State (3).

$$\pi(3)(1-\alpha_0) = \pi(1|3)(\alpha_1+\alpha_2) + \pi(E)\alpha_3 + \pi(3|1)\alpha_1$$

Using the relations we already obtained, we substitute $\pi(1|3)$ and $\pi(3|1)$:

$$\pi(3)(1-\alpha_0) = \pi(1)(\alpha_1+\alpha_2)\frac{\alpha_3}{1-\alpha_0} + \pi(E)\alpha_3 + \pi(3)\alpha_1\frac{\alpha_1}{1-\alpha_0}$$

One can check with some simple algebraic manipulations of these equations, that the solution we propose in Eq. 1 is the solution of the balance equations.

$$\begin{aligned} \pi(x|y) &= \pi(x)\alpha_y/(1-\alpha_0) \\ \pi(1) &= \pi(E)\alpha_1/(\alpha_1+\alpha_2) \\ \pi(2) &= \pi(E)\alpha_2/(\alpha_1+\alpha_2+\alpha_3) \\ \pi(3) &= \pi(E)\alpha_3/(\alpha_2+\alpha_3+\alpha_4) \\ \pi(4) &= \pi(E)\alpha_4/(\alpha_3+\alpha_4) \end{aligned} \qquad (1)$$

and $\pi(E)$ is obtained by normalization.

Note that this solution is the solution proved by Moyal et al. in [8] for a general matching model without loops on the matching graph:

$$\pi(w_1|..|w_k) = C\prod_{i=1}^{k} \frac{\alpha(w(i))}{\alpha(\Gamma(w(1),...,w(i)))} \qquad (2)$$

Thus one may expect that the multiplicative solution for the steady-state still holds under our assumptions.

4   Steady-State Distribution

We now prove that the Markov chain associated with any matching graph which has a loop on all nodes has a steady state distribution which has a multiplicative form. To prove the theorem, we use the following characterization of the steady-state distribution.

Property 1. *Consider a Markov chain with state space $\mathcal{E}$ and transition matrix P. Let π be a finite measure (i.e. $\|\pi\| < \infty$). If matrix Q defined by for all i and j in $\mathcal{E}$*

$$\pi(i)P[i,j] = \pi(j)Q[j,i]$$

is a stochastic matrix, then the steady state distribution of the Markov chain associated with P is obtained through normalization of π (i.e. $\pi/\|\pi\|$).

Remark 2. *If $P = Q$, the chain is reversible.*

Assuming that the multiplicative solution already known for matching graphs without loop still holds, one can formally obtain matrix Q and check if this matrix is stochastic. This is the key idea for the proof.

Remark 3. *As for all x we have $P[x,x] = \alpha_0$, then, by construction, $Q[x,x] = \alpha_0$.*

4.1   P4 Revisited

Consider again the example of the Markov chain associated with a P4 matching graph. We compute Q^t (instead of Q to simplify the presentation) by computing the product with the probabilities $\pi()$ given in Eq. 1. For instance $Q[i,i] = \alpha_0$ for all i, and:

$$Q[(1|4),(1)] = \frac{\pi((1))}{\pi((1|4))}P[(1),(1|4)] = \frac{1-\alpha_0}{\alpha_4}\alpha_4 = 1-\alpha_0$$

We denote $\beta_0 = 1 - \alpha_0$ to simplify the matrix formulation.

$$
Q^t =
\begin{bmatrix}
\alpha_0 & \alpha_1+\alpha_2 & \alpha_1+\alpha_2+\alpha_3 & \alpha_2+\alpha_3+\alpha_4 & \alpha_3+\alpha_4 & 0 & 0 & 0 & 0 & 0 & 0 \\
\alpha_1 & \alpha_0 & 0 & 0 & 0 & \beta_0 & 0 & \beta_0 & 0 & 0 & 0 \\
\alpha_2 & 0 & \alpha_0 & 0 & 0 & 0 & 0 & 0 & 0 & \beta_0 & 0 \\
\alpha_3 & 0 & 0 & \alpha_0 & 0 & 0 & 0 & 0 & \beta_0 & 0 & 0 \\
\alpha_4 & 0 & 0 & 0 & \alpha_0 & 0 & \beta_0 & 0 & 0 & 0 & \beta_0 \\
0 & \frac{\alpha_3(\alpha_3+\alpha_4)}{\beta_0} & 0 & 0 & \frac{\alpha_1(\alpha_3+\alpha_4)}{\beta_0} & \alpha_0 & 0 & 0 & 0 & 0 & 0 \\
0 & \frac{\alpha_4(\alpha_1+\alpha_2)}{\beta_0} & 0 & 0 & \frac{\alpha_1(\alpha_1+\alpha_2)}{\beta_0} & 0 & \alpha_0 & 0 & 0 & 0 & 0 \\
0 & \frac{\alpha_3(\alpha_3+\alpha_4)}{\beta_0} & 0 & \frac{\alpha_1(\alpha_2+\alpha_3+\alpha_4)}{\beta_0} & 0 & 0 & 0 & \alpha_0 & 0 & 0 & 0 \\
0 & \frac{\alpha_3(\alpha_1+\alpha_2)}{\beta_0} & 0 & \frac{\alpha_1^2}{\beta_0} & 0 & 0 & 0 & 0 & \alpha_0 & 0 & 0 \\
0 & 0 & \frac{\alpha_4^2}{\beta_0} & 0 & \frac{\alpha_2(\alpha_3+\alpha_4)}{\beta_0} & 0 & 0 & 0 & 0 & \alpha_0 & 0 \\
0 & 0 & \frac{\alpha_4(\alpha_1+\alpha_2+\alpha_3)}{\beta_0} & 0 & \frac{\alpha_2(\alpha_1+\alpha_2)}{\beta_0} & 0 & 0 & 0 & 0 & 0 & \alpha_0
\end{bmatrix}
$$

Clearly Q^t is column stochastic, thus Q is a stochastic matrix. And the result (i.e. Eq. 1) holds.

4.2   Main Result

Let us now proceed with our main result on the steady-state distribution of the Markov chain associated with the matching graph. We first state the result. Before proceeding with the proof, we give some technical lemmas for the 3 types of nodes, as mentioned during the analysis of the $P4$ example.

Theorem 1. *Let G be a graph with a loop on each vertex. Let n be the number of nodes of G. Let $\alpha_0, \alpha_1, ..., \alpha_n$ be a proper distribution of probability. The steady-state distribution of the Markov chain associated with matching graph G has a multiplicative form:*

$$\pi(m) = C \frac{\prod_{\psi=1}^{n} \alpha_{m(\psi)}}{\prod_{\psi=1}^{n} \alpha(\Gamma(Pre(m, \psi)))}.$$

where C is a normalization constant equal to $\pi(E)$.

Corollary 1. *Assume that state $(m|x)$ exists, then we have:*

$$\pi(m|x) = \pi(m) \frac{\alpha_x}{\alpha(\Gamma(Pre(m|x)))}.$$

The proof of the theorem is based on the analysis of matrix Q for the three types of node. First we need to study the graph properties of the Markov chain. We begin with the description of the edges.

Lemma 1. *A state m of the Markov chain a positive number of letters (i.e. $|m| > 0$) only has transitions to itself (because of null arrival event) and to states with size $|m| + 1$, or $|m| - 1$.*

Proof: It is a clear consequence of the description of the effect of an arrival.

Lemma 2. *A state m of the Markov chain with a positive number of letters (i.e. $|m| > 0$) has only one transition to a state with size $|m| - 1$ and the loop transition with probability α_0. The other transitions lead to states with size $|m|+1$ Furthermore, if m is not a maximal independent set, for all the letters x such that $(m|x)$ exists, we have $P(m, m|x) = \alpha_x$.*

Proof: First, let m be a state with a positive number of letter. We can write $m = (j|x)$, where j is a word (one can have $j = E$)). There exists a unique transition from j to $(j|x)$ with rate α_x due to the arrival of a letter x and the FCFM matching discipline. Indeed, the only possibility to increase the size of the state is the arrival of a letter which must be the last one due to the FCFM discipline.

Finally according to the previous lemma, all the remaining transitions leads to states with one more letter. And if $(m|x)$ exists, the only transition from m to $(m|x)$ is the arrival of letter x and it has probability α_x.

Lemma 3. *[Global balance equation for a maximal independent set] Assume that the size of the maximal independent sets is at least 2. A state is a maximal independent set if it is a word $(m|x)$ where m is also an independent set which does not contain an x. The only transition (except the loop) entering such a state comes from state m and has probability α_x. There exist at least two outgoing transitions: the loop with rate α_0 and all the transitions provoked by arrivals which delete one letter in $(m|x)$. Thus the global balance equation for such a state is:*

$$(1 - \alpha_0)\pi(m|x) = \pi(m)\alpha_x$$

Proof: Clearly, it exits a transition with probability α_x going from m to $(m|x)$ if x does not match with a letter of m. According to Lemma 1, the states which precede $(m|x)$ have size $|m|$, $|m| + 1$ (due to the loop) or $|m| + 2$. Clearly, the only transition entering $(m|x)$ from a state with a smaller number of letters is the transition going from m due to the FCFM discipline.

Let us prove now by contradiction that there does not exist any transition from a state $(m'|x)$ to a state $(m|x)$ with $|m'| = |m| + 2$. Assume that it is possible, then state $(m'|x)$ has a size larger than the size of $(m|x)$ and as an arrival provokes a transition from $(m'|x)$ to $(m|x)$, all the letters of $(m|x)$ are also in $(m'|x)$. Thus $(m|x)$ cannot be a maximal independent set as it is contained in a larger independent set.

Finally, state $(m|x)$ has an output degree which is at least 3. Indeed there exists a transition from $(m|x)$ to m since the arrival of x deletes the last letter x in the word (this is the effect of the loop on x in the matching graph). Furthermore, the arrivals of all the letters in m delete a letter of m and to not delete x. Therefore they provoke the transition to $(m'|x)$ with $|m'| = |m| - 1$. Finally we add the loop and the output degree is at least 3. The global balance equation is now trivial.

Property 2. *Let m be a state of the chain and x an arbitrary letter which is in $V \setminus \Gamma(m)$. We define the following subset of states:*

$$\Gamma^{-x}(m) = \{p \mid |p| = 1 + |m|, \ and \ p = m + x\}$$

Intuitively, the arrival of letter x in state p provokes a transition from p to m because letter x is deleted. Such a subset is only defined when x is in $V \setminus \Gamma(m)$. Indeed, if the word has size $|m| + 1$, it means that letter x does not interact with word m. Furthermore the cardinal of $\Gamma^{-x}(m)$ is $|m| + 1$. Indeed, one can add letter x anywhere inside word m.

Lemma 4. *Let j be a state (i.e. a word) of size n. For all letter x in $V \setminus \Gamma(j)$, we have:*

$$\sum_{l \in \Gamma^{-x}(j)} Q(j, l) = \alpha_x.$$

As the proof of this lemma is technical, it is postponed after the proof of the main theorem.

Proof of the Theorem Let us now proceed with the proof the main theorem. Remember we partition the states of the Markov chain into three subsets:

1. Empty state E
2. Maximal Independent states
3. Other states

We make the proof for the three types of states:

- Empty State E: for all letter x we have by construction:

$$Q(E, x) = P(x, E)\frac{\pi(x)}{\pi(E)}$$

From the definition of π in Theorem 1, $\pi(x) = \frac{\pi(E)\alpha_x}{\alpha(\Gamma(x))}$ and $P(x, E) = \alpha(\Gamma(x))$. After simplification,

$$Q(E, x) = \alpha_x$$

Thus $\sum_{x \in V} Q(E, x) = \sum_{x \in V} \alpha_x = 1 - \alpha_0$ and $Q(j, j) = \alpha_0$. Therefore the row sum is equal to 1 for state E.

- Maximal Independent set: Let p be a maximal independent set. According to Lemma 3, there exists now only one entry in the column associated with p in P. Thus there exists only one entry in the row associated with p in Q. Let m be this predecessor of p and let x the letter appended to m to obtain p (i.e. $p = (m|x)$). By construction:

$$Q(p, m) = P(m, p)\pi(m)/\pi(p).$$

From Lemma 3, $P(m, p) = \alpha_x$, and $(1 - \alpha_0)\pi(p) = \alpha_x\pi(m)$. Thus, $Q(p, m) = 1 - \alpha_0$. Furthermore $Q(p, p) = \alpha_0$. And all other entries of matrix Q for row p are 0. Therefore the row sum is equal to 1 for a state which is a maximal Independent set.

- Other states: we know due to Lemma 2 that, if it is not a maximal independent set, a state j of size n has one predecessor with size $n-1$, several predecessors of size $n+1$ and itself (with probability α_0). We will consider these three sets of predecessors in a separate way.

  - Let m the predecessor of size $n-1$. Let x the letter which has been appended (i.e. $j = m|x$). From Lemma 2 we have: $P(m, j) = \alpha_x$. From Corollary 1, we have

    $$\pi(j) = \pi(m|x) = \frac{\pi(m)\alpha_x}{\alpha(\Gamma(m|x))}$$

    Thus

    $$Q(j, m) = \alpha(\Gamma(m|x)) = \alpha(\Gamma(j))$$

  - We now consider the predecessors of j which have size $n+1$. Let $H(j)$ be this set. We will partition this set of states according to the letter x which provokes the transition. Thus,

    $$H(j) = \cup_{x \in V \setminus \Gamma(j)} \Gamma^{-x}(j).$$

As subsets $\Gamma^{-x}(j)$ do not intersect, this is a true partition. Thus,

$$\sum_{l \in H(j)} Q(j,l) = \sum_{x \in V \setminus \Gamma(j)} \sum_{l \in \Gamma^{-x}(j)} Q(j,l)$$

Technical Lemma 4 states that: $\sum_{l \in \Gamma^{-x}(j)} Q(j,l) = \alpha_x$. Thus,

$$\sum_{l \in H(j)} Q(j,l) = \sum_{x \in V \setminus \Gamma(j)} \alpha_x.$$

Combining both results, we get:

$$\sum_l Q(j,l) = \alpha_0 + \alpha(\Gamma(j)) + \sum_{l \in H(j)} Q(j,l) = \alpha_0 + \alpha(\Gamma(j)) + \sum_{x \in V \setminus \Gamma(j)} \alpha_x = \sum_{x \in V} \alpha_x = 1$$

And the proof is complete.

4.3   Proof of the Technical Lemma

We want to prove that for an arbitrary word j of size n and for all letter x in $V \setminus \Gamma(j)$, we have:

$$\sum_{l \in \Gamma^{-x}(j)} Q(j,l) = \alpha_x.$$

Let us first explain the terms involved in the summation and the way we combine them. By definition we have:

$$Q(j,l) = P(l,j)\pi(l)/\pi(j).$$

and by assumptions, the solution is:

$$\pi(l) = \pi(E)\frac{\prod_{\psi=1}^{n+1} \alpha_{l(\psi)}}{\prod_{\psi=1}^{n+1} \alpha(\Gamma(Pre(l,\psi)))}.$$

As $l \in \Gamma^{-x}(j)$, one can write $l = j + x$ and

$$\prod_{\psi=1}^{n+1} \alpha_{l(\psi)} = \alpha_x \prod_{\psi=1}^{n} \alpha_{j(\psi)}$$

Let us now study the denominator. Consider an arbitrary word $(j + x)$. There exists an index ψ (which can be 0) such that this word is $(Pre(j,\psi)|x|Suf(j, n - \psi))$. This formulation allows to obtain the transition probability and the steady-state distribution for all the values of ψ and obtain an induction on partial sums on ψ

Let us begin the induction with $\psi = 0$ associated with $Q(j,(x|j))$. Clearly,

$$P((x|j),j) = \alpha(\Gamma(x)).$$

Now the denominator of $\pi(x|j)$ is according to the assumptions equal to $\prod_{\psi=1}^{n+1} \alpha(\Gamma(Pre((x|j),\psi)))$. It is easy to remark that:

$$\prod_{\psi=1}^{n+1} \alpha(\Gamma(Pre((x|j),\psi))) = \prod_{\psi=0}^{n} \alpha(\Gamma(x|Pre(j,\psi)))$$

This remark is used to simplify the factorization.

$$\pi(x|j) = \pi(E)\alpha_x \frac{\prod_{\psi=1}^{n} \alpha_{j(\psi)}}{\prod_{\psi=0}^{n} \alpha(\Gamma((x|Pre(j,\psi))))}$$

and,

$$\pi(j) = \pi(E)\frac{\prod_{\psi=1}^{n} \alpha_{j(\psi)}}{\prod_{\psi=1}^{n} \alpha(\Gamma(Pre(j,\psi)))}$$

and finally after cancellation of terms:

$$Q(j,(x|j)) = \alpha_x \alpha(\Gamma(x))\frac{\prod_{\psi=1}^{n} \alpha(\Gamma(Pre(j,\psi)))}{\prod_{\psi=0}^{n} \alpha(\Gamma((x|Pre(j,\psi))))}$$

We now have to compute the term associated with $\psi = 1$ (remember that we want to make an induction on partial sums). The state of the chain is $(Pre(j,1)|x|Suf(j,n-1))$. The transition from this state to j is provoked by the arrival of letters which match with x but which are not matched with the first letter of this word due to the FCFM matching discipline. More formally:

$$P((Pre(j,1)|x|Suf(j,n-1)),j) = \alpha(\Gamma(x) \setminus \Gamma(Pre(j,1))),$$

and

$$\pi((Pre(j,1)|x|Suf(j,n-1))) = \pi(E)\alpha_x \frac{\prod_{\psi=1}^{n} \alpha_{j(\psi)}}{\alpha(\Gamma(Pre(j,1))) \prod_{\psi=1}^{n} \alpha(\Gamma((x|Pre(j,\psi))))},$$

and finally after substation and cancellation, we get:

$$Q(j,(Pre(j,1)|x|\ Suf(j,n-1))) =$$
$$\alpha_x \alpha(\Gamma(x) \setminus \Gamma(Pre(j,1)))\frac{\prod_{\psi=1}^{n} \alpha(\Gamma(Pre(j,\psi)))}{\alpha(\Gamma(Pre(j,1))) \prod_{\psi=1}^{n} \alpha(\Gamma((x|Pre(j,\psi))))}.$$

We now compute the sum of these first two elements. After factorization:

$$Q(j,(x|j)) + Q(j,(\ Pre(j,1)|x|Suf(j,n-1))) =$$
$$\alpha_x \frac{\prod_{\psi=1}^{n} \alpha(\Gamma(Pre(l,\psi)))}{\prod_{\psi=1}^{n} \alpha(\Gamma((x|Pre(j,\psi))))}\left[\frac{\alpha(\Gamma(x))}{\alpha(\Gamma((x|Pre(j,0))))} + \frac{\alpha(\Gamma(x) \setminus \Gamma(Pre(j,1)))}{\alpha(\Gamma(Pre(j,1)))}\right].$$

As $(Pre(j,0))$ is the empty word, we have $\Gamma((x|Pre(j,0)))) = \Gamma(x)$, and the first part of the summation simplifies.

$$Q(j,(x|j)) + Q(j,(Pre(j,1)|x|\ Suf(j,n-1))) =$$
$$\alpha_x \frac{\prod_{\psi=1}^{n} \alpha(\Gamma(Pre(j,\psi)))}{\prod_{\psi=1}^{n} \alpha(\Gamma((x|Pre(j,\psi))))}\left[1 + \frac{\alpha(\Gamma(x) \setminus \Gamma(Pre(j,1)))}{\alpha(\Gamma(Pre(j,1)))}\right].$$

Thus:

$$Q(j, (x|j)) + Q(j, (Pre(j, 1)|x|\ Suf(j, n-1))) =$$

$$\alpha_x \frac{\prod_{\psi=1}^{n} \alpha(\Gamma(Pre(l, \psi)))}{\prod_{\psi=1}^{n} \alpha(\Gamma((x|Pre(j, \psi))))} \frac{\alpha(\Gamma(Pre(j,1))) + \alpha(\Gamma(x) \backslash \Gamma(Pre(j,1)))}{\alpha(\Gamma(Pre(j,1)))}.$$

For all words a and b, we have:

$$\alpha(\Gamma(a) \backslash \Gamma(b)) + \alpha(\Gamma(b)) = \alpha(\Gamma(a|b)).$$

Using $a = x$ and $b = Pre(j, 1)$, after substitution we get:

$$Q(j, (x|j)) + Q(j, (Pre(j, 1)|x|\ Suf(j, n-1))) =$$

$$\alpha_x \frac{\prod_{\psi=1}^{n} \alpha(\Gamma(Pre(j, \psi)))}{\prod_{\psi=1}^{n} \alpha(\Gamma((x|Pre(j, \psi))))} \frac{\alpha(\Gamma((x|Pre(j,1))))}{\alpha(\Gamma(Pre(j,1)))}.$$

$\alpha(\Gamma((x|Pr(j, 1))))$ is in the numerator and the denominator. We cancel this term and we get:

$$Q(j, (x|j)) + Q(j, (Pre(j, 1)|x|\ Suf(j, n-1))) =$$

$$\alpha_x \frac{\prod_{\psi=1}^{n} \alpha(\Gamma(Pre(j, \psi)))}{\alpha(\Gamma(Pre(j,1))) \prod_{\psi=2}^{n} \alpha(\Gamma((x|Pre(j, \psi))))}.$$

Let us now consider the induction on the number of terms we add. Assume that we consider them and accumulate them according to the index of x in the word. Assume that we have proved that for all ψ between 0 and $\nu - 1$, we have stated that the summation of the first ν terms is equal to:

$$\frac{\alpha_x \prod_{\psi=1}^{n} \alpha(\Gamma(Pre(j, \psi)))}{\prod_{\psi=1}^{\nu-1} \alpha(\Gamma(Pre(j, \psi))) \prod_{\psi=\nu}^{n} \alpha(\Gamma((x|Pre(j, \psi))))}.$$

We now consider term with index ν. We have to compute $Q(j, (Pre(j, \nu)|x|\ Suf(j, n - \nu)))$ and add it to the previous partial sum. As usual due to the FCFM matching discipline

$$P((Pre(j, \nu)|x|Suf(j, n - \nu)), j) = \alpha(\Gamma(x) \backslash \Gamma(Pre(j, \nu))),$$

By assumption on the multiplicative solution for the steady-state:

$$\pi((Pre(j, \nu)|x|\ Suf(j, n - \nu))) =$$

$$\pi(E) \frac{\alpha_x \prod_{\psi=1}^{n} \alpha_{j(\psi)}}{\prod_{\psi=1}^{\nu} \alpha(\Gamma(Pre(j, \psi))) \prod_{\psi=\nu}^{n} \alpha(\Gamma((x|Pre(j, \psi))))}.$$

Thus,

$$Q(j, (Pre(j, \nu)|x|\ Suf(j, n - \nu))) =$$

$$\alpha(\Gamma(x) \backslash \Gamma(Pr(j, \nu))) \frac{\alpha_x \prod_{\psi=1}^{n} \alpha(\Gamma(Pre(j, \psi)))}{\prod_{\psi=1}^{\nu} \alpha(\Gamma(Pre(j, \psi))) \prod_{\psi=\nu}^{n} \alpha(\Gamma((x|Pre(j, \psi))))}.$$

After summation with the previous partial sum (given by the induction assumption) and factorization, we get that the new partial summation is equal to:

$$\frac{\alpha_x \prod_{\psi=1}^{n} \alpha(\Gamma(P(j,\psi)))}{\prod_{\psi=1}^{\nu-1} \alpha(\Gamma(Pre(j,\psi))) \prod_{\psi=\nu}^{n} \alpha(\Gamma((x|Pre(j,\psi)))} \left[1 + \frac{\alpha(\Gamma(x) \setminus \Gamma(Pre(j,\nu)))}{\alpha(\Gamma(Pre(j,\nu)))} \right].$$

We use a similar argument to simplify:

$$1 + \frac{\alpha(\Gamma(x) - \Gamma(Pre(j,\nu)))}{\alpha(\Gamma(Pre(j,\nu)))} = \frac{\alpha(\Gamma((x|Pre(j,\nu))))}{\alpha(\Gamma(Pre(j,\nu)))}.$$

After substitution the sum is equal to:

$$\frac{\alpha_x \prod_{\psi=1}^{n} \alpha(\Gamma(Pre(j,\psi)))}{\prod_{\psi=1}^{\nu-1} \alpha(\Gamma(Pre(j,\psi))) \prod_{\psi=\nu}^{n} \alpha(\Gamma((x|Pre(j,\psi))))} \; \frac{\alpha(\Gamma((x|Pre(j,\nu))))}{\alpha(\Gamma(Pre(j,\nu)))} =$$

$$\frac{\alpha_x \prod_{\psi=1}^{n} \alpha(\Gamma(Pre(j,\psi)))}{\prod_{\psi=1}^{\nu} \alpha(\Gamma(Pre(j,\psi))) \prod_{\psi=\nu+1}^{n} \alpha(\Gamma((x|Pre(j,\psi))))}.$$

Thus the induction holds. Now let us compute the sum of all the elements for ψ between 0 and n. According to the induction assumptions which is now proved, this sum is equal to

$$\alpha_x \frac{\prod_{\psi=1}^{n} \alpha(\Gamma(Pre(j,\psi)))}{\prod_{\psi=1}^{n} \alpha(\Gamma(Pre(j,\psi)))} = \alpha_x,$$

and the proof of the Lemma is complete.

5   Conclusions and Remarks

This paper is a sequel of [9] where we proved that adding a new edge in a matching graph without loops may lead to a performance paradox: the expectation of the total number of customers increase after the addition of the edge. See also [17] for an extended version of this paper.

Our aim was to prove or disprove the existence of the same paradox for matching graph with loops. The first step was to prove that the steady state solution has a multiplicative form. However even with this result, the existence of a paradox similar to the one shown in [9] is still an open problem as all the examples studied so far do not exhibit the same paradox we found in [9]. Note however that the chains we obtain with this new model are all finite while the chains studied in [17] are infinite.

References

1. Mairesse, J., Moyal, P.: Stability of the stochastic matching model. J. Appl. Probab. **53**(4), 1064–1077 (2018)

2. Gelenbe, E.: Product-form queuing networks with negative and positive customers. J. Appl. Probab. **28**, 656–663 (1991)
3. Fourneau, J.-M., Gelenbe, E., Suros, R.: G-networks with multiple classes of positive and negative customers. Theoret. Comput. Sci. **155**, 141–156 (1996)
4. Dao-Thi, T.-H., Fourneau, J.-M., Tran, M.-A.: Network of queues with inert customers and signals. In: 7th International Conference on Performance Evaluation Methodologies and Tools, ValueTools 2013, Italy, pp. 155–164. ICST/ACM (2013)
5. Marsan, M.A., Balbo, G., Bobbio, A., Chiola, G., Conte, G., Cumani, A.: On Petri nets with stochastic timing. In: International Workshop on Timed Petri Nets, Torino, Italy, 1–3 July 1985, pp. 80–87. IEEE Computer Society (1985)
6. Lazar, A.A., Robertazzi, T.G.: Markovian Petri net protocols with product form solution. Perform. Eval. **12**(1), 67–77 (1991)
7. Haddad, S., Moreaux, P., Sereno, M., Silva, M.: Product-form and stochastic Petri nets: a structural approach. Perform. Eval. **59**, 313–336 (2005)
8. Moyal, P., Bušić, A., Mairesse, J.: A product form for the general stochastic matching model. J. Appl. Probab. **57**(2), 449–468 (2021)
9. Cadas, A., Doncel, J., Fourneau, J.-M., Busic, A.: Flexibility can hurt dynamic matching system performance. ACM SIGMETRICS Perform. Eval. Rev. **49**(3), 37–42 (2021). IFIP Performance Evaluation (Short Paper)
10. Bušić, A., Gupta, V., Mairesse, J.: Stability of the bipartite matching model. Adv. Appl. Probab. **45**(2), 351–378 (2013)
11. Caldentey, R., Kaplan, E.H., Weiss, G.: FCFS infinite bipartite matching of servers and customers. Adv. Appl. Probab. **41**(3), 695730 (2009)
12. United Network for Organ Sharing. https://unos.org/wp-content/uploads/unos/living_donation_kidneypaired.pdf
13. Unver, U.: Dynamic kidney exchange. Rev. Econ. Stud. **77**(1), 372–414 (2010)
14. Ashlagi, I., Jaillet, P., Manshadi, V.H.: Kidney exchange in dynamic sparse heterogenous pools. In: Proceedings of the Fourteenth ACM Conference on Electronic Commerce, EC 2013, pp. 25–26. ACM, New York (2013)
15. Begeot, J., Marcovici, I., Moyal, P., Rahme, Y.: A general stochastic matching model on multigraphs (2020). arXiv preprint https://arxiv.org/abs/2011.05169
16. Fourneau, J.M., Mahjoub, Y.A.E., Quessette, F., Vekris, D.: XBorne 2016: a brief introduction. In: Czachórski, T., Gelenbe, E., Grochla, K., Lent, R. (eds.) ISCIS 2016. CCIS, vol. 659, pp. 134–141. Springer, Cham (2016). https://doi.org/10.1007/978-3-319-47217-1_15
17. Cadas, A., Doncel, J., Fourneau, J.-M., Busic, A.: Flexibility can hurt dynamic matching system performance, extended version on arXiv (2020)

$M/M/c$/Setup Queues: Conditional Mean Waiting Times and a Loop Algorithm to Derive Customer Equilibrium Threshold Strategy

Hung Q. Nguyen[1]([✉]) [iD] and Tuan Phung-Duc[2] [iD]

[1] Graduate School of Science and Technology, University of Tsukuba,
1-1-1 Tennodai, Tsukuba, Ibaraki 305-8573, Japan
nguyen.quoc.hung.xu@alumni.tsukuba.ac.jp
[2] Institute of Systems and Information Engineering, University of Tsukuba,
1-1-1 Tennodai, Tsukuba, Ibaraki 305-8573, Japan
tuan@sk.tsukuba.ac.jp

Abstract. This paper considers an $M/M/c$/Setup queueing system motivated from data centers under an ON–OFF policy (that is, with idle servers turned off to save energy). We derive and show features of the expected waiting times of customers at every system state. While the system can be modeled using only two variables, it is necessary to adopt one more dimension to derive the expected waiting time at a specific state. Furthermore, we propose a loop algorithm to compute the equilibrium threshold strategy of strategic customers, and calculate performance measures of the system in equilibrium. Finally, we consider optimal system design through several numerical examples, focusing on maximizing social welfare.

Keywords: Queueing system · Setup time · Game theory · Equilibrium threshold strategy · Data center

1 Introduction

Large-scale data centers have become indispensable infrastructure for today's information society. These data centers provide core support for various applications such as social networking services (SNS), video-conferencing, storage services, etc. In data centers, a huge number of servers are available but they are not all processing jobs all the time. It is reported that the server utilization in data centers is merely 30–40% (Barroso and Hölzle 2007). Even an idle server, however, still consumes about 60–70% of its active-state energy even when not processing a job (Barroso and Hölzle 2007; Gandhi et al. 2014). This poses the question of whether we should reduce the number of servers in data centers. However, because Internet traffic is stochastic, too few servers results in a very

K. Gilly and N. Thomas (Eds.): EPEW 2022, LNCS 13659, pp. 86–99, 2023.
https://doi.org/10.1007/978-3-031-25049-1_6

long response time during peak periods, and a sufficient number to fully handle peak traffic leaves most of them idle, and hence wasting energy, off-peak.

A natural and simple solution is to turn servers on and off dynamically, according to the number of available jobs. However, a server needs a setup time to become active and serve a job. During the setup time, the server consumes as much energy as when it is active but cannot process a job. Thus, while this kind of on-off control may reduce energy consumption, it may also extend the response time. Data centers with on-off control can be modeled using queues with setup time.

Queues with setup time have attracted considerable attention because these models are motivated from power conservation in data centers as presented above, as well as 5G systems where resource is scaled in and out according to traffic demand (Gandhi et al. 2014; Phung-Duc 2017; Hsieh et al. 2022). Well-regulated on-off control is expected to save energy consumption while keeping the response time reasonably fast. (Mitrani 2013) investigated optimal control of service centers in which servers are turned on and turned off in a block. (Mitrani 2011) considered optimal control when customers are impatient and their presence in the queue expires after an exponentially distributed random time.

An interesting feature of queues with setup time is the response time distribution, which is not obtained directly from the queue length distribution as it is, via Little's law (Keilson and Servi 1988), in other queueing systems. The reason is that the response time of a tagged customer depends not only on the state which the customer sees upon arrival, but also on future arrivals behind him during the waiting time. This problem is studied in (Phung-Duc and Kawanishi 2020).

In this paper, we further study the model from a strategic queueing perspective where customers are rational under decision theory. Queues with strategic customers have been extensively studied. In the 1-dimensional problem (Naor 1969), and most studies following this line of research (summarized in (Hassin and Haviv 2003)), customers follow a threshold strategy based on the only dimension of the system state that they observe, and later customer arrivals do not affect the waiting times of those who are already enqueued. As mentioned above, customer waiting time depends on both the state upon arrival, and also on future arrivals. Furthermore, future customers are also strategic, which poses a problem more complicated and challenging, and, it would appear, not yet studied in the literature. The main contributions of this study can be summarized as follows:

- Conditional expected customer waiting times are derived recursively with respect to three variables in the non-strategic queueing scenario, and four variables in the case where customers are strategic.
- A loop algorithm is proposed to compute the equilibrium threshold strategy.
- Optimizations, considering social welfare that balances the benefits of both demand and supply sides, are conducted through numerical examples.

The rest of this paper is organized as follows. Section 2 provides descriptions of the model. Section 3 derives expected waiting times at each system state in the non-strategic queueing scenario. Section 4 considers the customers' strategic behaviors, and derives the system equilibrium. In Sect. 5, several performance measures of the equilibrium system are derived. Section 6 illustrates the results through numerical examples. Finally, we conclude the paper to propose several suggestions for future research.

2   Model Descriptions

Consider a data center with c identical servers controlled under the following policy:

1. An active, i.e. ON, server is turned OFF immediately if there is no waiting customer; in other words, there is no idle server at any time.
2. By nature, it takes time for an OFF server to start up, and a server is in a SETUP mode during the startup time.
3. An OFF server is turned ON upon arrival of a customer if the number of customers is larger than the total number of ON servers and SETUP servers.

We assume that the server setup times follow an exponential distribution with rate α.

Assume that customers arrive according to a Poisson process with rate λ, and that service times follow an exponential distribution with rate μ. Let R and C, respectively, denote the customers' identical service value and waiting cost rate. Denote by C_a and C_s the power consumption cost rates of ON (i.e. active) servers and SETUP servers, respectively.

3   Conditional Expected Waiting Times

Denote by $T(n, i, j)$, where $(n, i, j) \in \{0, 1, ...\} \times \{0, 1, ..., c\} \times \{0, 1, ...\}$, the expected waiting time of a customer at a position n, observing i ON servers, and with j customers behind. A position 0 indicates that the tagged customer is currently in service. The number of servers in the SETUP state is $\min(n+j, c-i)$.

We immediately obtain $T(0, i, j) = 0$. For $n \geq 1$, according to the first-hitting-time analysis, the expected waiting time is recursively calculated as

$$T(n, i, j) = \frac{1}{\lambda + i\mu + \min(n+j, c-i)\alpha} + \frac{\lambda}{\lambda + i\mu + \min(n+j, c-i)\alpha} T(n, i, j+1)$$

$$+ \frac{i\mu}{\lambda + i\mu + \min(n+j, c-i)\alpha} T(n-1, i, j)$$

$$+ \frac{\min(n+j, c-i)\alpha}{\lambda + i\mu + \min(n+j, c-i)\alpha} T(n-1, i+1, j). \tag{1}$$

To keep the formula simple, we temporarily assume that $T(n, -1, j)$ and $T(n, c+1, j)$ make sense and take an arbitrary constant value, which does not affect the overall result because the weight attached to those amounts is always 0.

Turning back to the formula (1), it can be seen that j keeps increasing to infinity in the recursion. However, the mean waiting time remains unchanged when j becomes large enough. To demonstrate this, we start with $n = 1$. Consider a tagged customer at an arbitrary state $(1, i, j)$ where $1 + i + j \geq c$. Under this condition, there are no further OFF servers left to start up, so any further customer arrivals will not affect the number of servers in the SETUP state, which remains at $c - i$. Consequently, for a fixed value of i, the distributions of the chains started at any state $(1, i, j)$ with $1 + i + j \geq c$ are the same and do not depend on j, as we can see in Fig. 1.

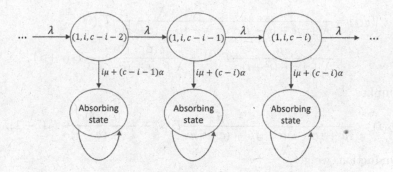

Fig. 1. Transition diagram for $n = 1$.

As a result, we obtain $T(1, i, j) = T(1, i, j+1) = T_1(i)$ for all i and j satisfying $1 + i + j \geq c$. Substituting into (1) yields

$$T_1(i) = \frac{1}{\lambda + i\mu + (c - i)\alpha} + \frac{\lambda}{\lambda + i\mu + (c - i)\alpha} T_1(i),$$

which implies

$$T_1(i) = \frac{1}{i\mu + (c - i)\alpha}.$$

In other words, $T(1, i, j) = \frac{1}{i\mu + (c-i)\alpha}$ for all i, j satisfying $1 + i + j \geq c$. For any i, j such that $1 + i + j < c$, $T(1, i, j)$ can be recursively calculated from (1).

Now let us turn to the case where $n = 2$. Consider a tagged customer at an arbitrary state $(2, i, j)$, where $1 + i + j \geq c$. As illustrated in *Figure* 2, for a fixed value of i, the chain distributions starting at any state $(2, i, j)$ with $1 + i + j \geq c$ are the same, since the distributions of the chains below them (at $n = 1$) are identical in pairs, as previously pointed out.

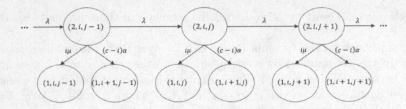

Fig. 2. Transition diagram for $n = 2$.

We thus obtain $T(2, i, j) = T(2, i, j+1) = T_2(i)$ for all i, j satisfying $1+i+j \geq c$. Substituting into (1) yields

$$T_2(i) = \frac{1}{\lambda + i\mu + (c-i)\alpha} + \frac{\lambda}{\lambda + i\mu + (c-i)\alpha} T_2(i)$$

$$+ \frac{i\mu}{\lambda + i\mu + (c-i)\alpha} T_1(i) + \frac{(c-i)\alpha}{\lambda + i\mu + (c-i)\alpha} T_1(i+1),$$

which implies

$$T_2(i) = \frac{1}{i\mu + (c-i)\alpha} + \frac{i\mu}{i\mu + (c-i)\alpha} T_1(i) + \frac{(c-i)\alpha}{i\mu + (c-i)\alpha} T_1(i+1).$$

By induction, we get

$$T_n(i) = \frac{1}{i\mu + (c-i)\alpha} + \frac{i\mu}{i\mu + (c-i)\alpha} T_{n-1}(i) + \frac{(c-i)\alpha}{i\mu + (c-i)\alpha} T_{n-1}(i+1).$$

for all i, j satisfying $1 + i + j \geq c$, and

$$T(n, c, j) = T_n(c) = \frac{n}{c\mu},$$

for all $n \geq 0$. For any n, i, j such that $n + i + j < c$, $T(n, i, j)$ can be recursively calculated from (1) with $T(n, c, j) = T_n(c)$ being the upper bound for the recursion.

Intuitively, this result means that, when the number of customers behind a tagged customer (at position n) becomes so large that no servers are turned OFF until the tagged customer gets to position 0, the expected waiting time of the tagged customer no longer depends on the number of customers behind him.

Remark 1 (Monotonicity properties of waiting times). $T(n, i, j)$ is non-decreasing in n and non-increasing in j.

Although the above monotonicities of waiting times are difficult to obtain through rigorous proof, they are intuitively straightforward and can be reinforced with a number of numerical examples. In what follows, we provide intuitive explanations of those properties.

- $T(n, i, j)$ is non-decreasing in n. This means that, ceteris paribus, the closer a tagged customer is to the servers, the sooner he can expect to be served.

- $T(n, i, j)$ is non-increasing in j. Unlike most basic queueing settings in which later comers do not affect the expected waiting time of customers already present in the queue, it is interesting to note that the waiting time of a tagged customer at a specific time is non-increasing with the number of customers queueing up behind him in the current model. This is because the more customers arrive after the tagged customer, the more OFF servers (if any) are turned ON; hence increasing the possibility of entering a server (which is currently in the SETUP mode) sooner. When all servers are ON or in SETUP mode, a customer arrival cannot cause any more servers to be turned ON, and also does not affect the waiting time of the tagged customer (as shown earlier). Furthermore, the tagged customer can move only toward the front of the queue over time (if he does not give up waiting, and leave).

4 Customer Behaviors

4.1 Preliminary

Assume that system states are observable to customers. Upon encountering a state (n, i, j), customers evaluate the expected utility of their joining decision, denoted $\mathsf{E}[U]$, by subtracting their waiting cost from the service value as follows.

$$\mathsf{E}[U] = R - C\left(T(n, i, j) + \frac{1}{\mu}\right).$$

At certain times, customers have two choices to make. Upon arriving, customers can choose between joining and balking the queue. Assume that the utility of "to balk" is zero.

Join or Balk? A tagged customer who arrives at position n and observes i active servers joins the system if

$$R - C\left(T(n, i, 0) + \frac{1}{\mu}\right) \geq 0,$$

and balks otherwise. From *Remark 1*, we have $T(n, i, 0) \leq T(n + 1, i, 0)$, which implies that there exists a threshold level of n, denoted n_i, corresponding to each fixed value of i such that $R - C\left(T(n, i, 0) + \frac{1}{\mu}\right) \geq 0$ and $R - C\left(T(n + 1, i, 0) + \frac{1}{\mu}\right) < 0$ (because it is intuitive that $T(n, i, 0) \to +\infty$ as $n \to +\infty$). We can then conclude that the joining strategy of customers is represented by an $(c + 1)$-dimensional vector $\eta = (n_0, n_1, ..., n_c)$ which encodes $c + 1$ threshold levels corresponding to the $c + 1$ possible cases of the number of active servers.

4.2   Equilibrium

In this section, we derive the equilibrium strategy of customers. Notice that, when customers adopt a threshold strategy $\eta = (n_0, n_1, ..., n_c)$, the recursive formula (1) also changes according to η. Therefore, we need to add η as a variable of the waiting time function, i.e., $T(n, i, j, \eta)$. It should be noted that the recursion in (1) is a special case where $T(n, i, j, (+\infty, +\infty, ..., +\infty))$ is derived. For an arbitrary η, the recursion becomes

$$T(n,i,j,\eta) = \begin{cases} \frac{1}{\lambda+i\mu+\min(n+j,c-i)\alpha} + \frac{\lambda}{\lambda+i\mu+\min(n+j,c-i)\alpha}T(n,i,j+1,\eta) \\ \quad + \frac{i\mu}{\lambda+i\mu+\min(n+j,c-i)\alpha}T(n-1,i,j,\eta) \\ \quad + \frac{\min(n+j,c-i)\alpha}{\lambda+i\mu+\min(n+j,c-i)\alpha}T(n-1,i+1,j,\eta) \quad \text{if } n+j < n_i \\ \frac{1}{i\mu+\min(n+j,c-i)\alpha} + \frac{i\mu}{i\mu+\min(n+j,c-i)\alpha}T(n-1,i,j,\eta) \\ \quad + \frac{\min(n+j,c-i)\alpha}{i\mu+\min(n+j,c-i)\alpha}T(n-1,i+1,j,\eta) \quad \text{if } n+j \geq n_i. \end{cases} \quad (2)$$

We are interested in deriving a universal equilibrium strategy, that is, a strategy adopted by all customers that no one finds an incentive to deviate from. To this end, we recall two definitions:

Definition 1 (Best response function). *The best response function $\mathcal{F}$ is defined as*

$$\mathcal{F} \colon \mathbb{N}^{c+1} \longrightarrow \mathbb{N}^{c+1}$$

$$\eta \longmapsto \nu,$$

where the vector ν encodes the threshold strategy of an individual customer given that the threshold strategy η is adopted by every customer.

Definition 2 (Equilibrium threshold strategy). *A threshold strategy η^* is called an equilibrium threshold strategy if it is the best response against itself. In other words,*

$$\eta^* = \mathcal{F}(\eta^*).$$

We now propose an algorithm to derive an equilibrium threshold strategy η^*.

Algorithm 1. Deriving the equilibrium threshold strategy

1: $\eta \leftarrow \eta_0 = \left(n_0^{(0)}, n_1^{(0)}, ..., n_c^{(0)}\right)$ ▷ arbitrary initial threshold strategy
2: Calculate waiting times $T(n, i, j, \eta)$ ▷ from equation (2)
3: $\nu \leftarrow \mathcal{F}(\eta)$
4: **while** $\eta \neq \nu$ **do**
5:     $\eta \leftarrow \nu$
6:     Calculate waiting times $T(n, i, j, \eta)$ ▷ from equation (2)
7:     $\nu \leftarrow \mathcal{F}(\eta)$
8: **end while**
9: **return** η
10: $\eta^* \leftarrow \eta$ ▷ equilibrium threshold strategy

---

In the first step of *Algorithm 1*, $\eta_0 = \left(n_0^{(0)}, n_1^{(0)}, ..., n_c^{(0)}\right)$ can be chosen arbitrarily. In the numerical examples shown later, we will see that this initial choice does not affect the equilibrium threshold strategy. Intuitively, this initialized vector reflects customers' initial belief in the strategy adopted by other customers involved in the system. As the system evolves, the information on the strategy is continually updated, which may make the best response to the strategy converge to the equilibrium strategy, and the Markov chain also reaches its steady-state distribution.

Note that the above algorithm is based on a premise that customers are assumed to always follow threshold strategies with respect to the position. To be more precise, it is assumed (and is reinforced by numerical examples) that the monotone property of the waiting time function with respect to the position is maintained in every loop of *Algorithm 1*. As such, in steps 3 and 7 of *Algorithm 1*, the best response is calculated by obtaining ν_i corresponding to each fixed value of i such that $R - C\left(T(\nu_i, i, 0, \eta) + \frac{1}{\mu}\right) \geq 0$ and $R - C\left(T(\nu_i + 1, i, 0, \eta) + \frac{1}{\mu}\right) < 0$.

5 Performance Measures

Several system performance measures, namely, mean number of jobs, mean number of customers, mean number of switches and power consumption, are derived in (Phung- Duc 2017) under a non-strategic queueing context. Below, we derive performance measures of the system in equilibrium, focusing on economic measures.

Let $L(t), I(t)$, respectively, denote the total number of customers and the number of active servers in the system at time t. Then, the process $\{(L(t), I(t)) \mid t \geq 0\}$ is a continuous-time Markov chain with the state space $\mathbb{S} = \{(l, i)\}$, where $i = 0, 1, ..., c$ and $l = i, i + 1, ...$. We then obtain the infinitesimal generator $\mathcal{Q}$ of the Markov chain modeling the system in equilibrium. The steady-state probabilities are defined as $\pi = (\pi_0, \pi_1, ..., \pi_c)$, where $\pi_i = (\pi_{i,i}, \pi_{i+1,i}, ...)$, is a vector encoding all probabilities when there are i ON servers at the steady state. These probabilities are obtained by solving the following equations:

$$\begin{cases} \pi \mathcal{Q} = \mathbf{0}, \\ \pi e = 1, \end{cases}$$

where $\mathbf{0}$ is a zero vector of appropriate length, and e is a vector of all entries 1 with appropriate length.

Similarly to the results in (Phung-Duc 2017), the mean number of customers, the mean number of servers in ON and SETUP modes, respectively denoted by $\mathsf{E}[L]$, $\mathsf{E}[A]$ and $\mathsf{E}[S]$, are given by

$$E[L] = \sum_l \sum_i l\pi_{l,i},$$

$$E[A] = \sum_l \sum_i i\pi_{l,i},$$

$$E[S] = \sum_l \sum_i \min\{c - i, l - i\}\pi_{l,i}.$$

The system's mean power consumption is given by

$$P = C_a E[A] + C_s E[S].$$

Social welfare is given by

$$SW = (1 - \xi)\lambda R - CE[L] - C_a E[A] - C_s E[S],$$

where ξ is the balking probability of customers obtained by taking the sum of all probabilities of the states at which the queue length is larger than or equal to the threshold.

If a fixed fee θ is imposed on each customer (this fee can be also interpreted as a *toll fee*, as in (Naor 1969)), social welfare can be further separated into two parts: customer welfare and the profit of the data center (here, we neglect fixed costs other than power consumption). Customer welfare, denoted by $SW^{(demand)}$, is given by

$$SW^{(demand)} = (1 - \xi)\lambda(R - \theta) - CE[L].$$

The profit obtained by the data center, denoted by $SW^{(supply)}$, is given by

$$SW^{(supply)} = (1 - \xi)\lambda\theta - C_a E[A] - C_s E[S].$$

6   Numerical Examples

In this section, we carry out numerical analyses in which controllable system and economic parameters are tuned for optimal design of the system. In all examples, $\lambda = 4$, and $C = 10$.

Example 1: Convergence of Algorithm 1 In the example in Fig. 3, we set $\mu = 1$, $\alpha = 8$, $c = 5$, $R = 30$, $C_a = 7$ and $C_s = 4$. It can be seen that all three possible initial values for η_0 make the algorithm to converge to the same equilibrium threshold strategy $\eta^* = (14, 13, 12, 11, 10, 10)$. When $\eta_0 = (0, 0, 0, 0, 0, 0)$ or $\eta_0 = (1, 2, 3, 4, 5, 6)$, it takes only one iteration for the algorithm to find the equilibrium, while two iterations are needed for $\eta_0 = (+\infty, +\infty, +\infty, +\infty, +\infty, +\infty)$.

In all following examples, the algorithm also always converges to a unique equilibrium threshold strategy. We will compute performance measures of the system based on the derived equilibrium.

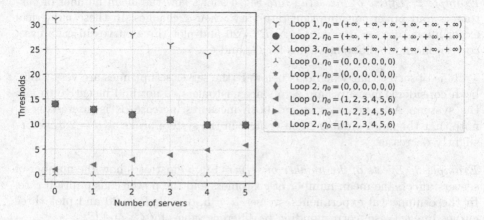

Fig. 3. Convergence to equilibrium strategy.

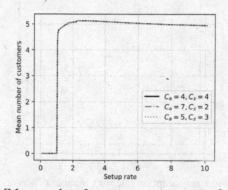

[Mean number of customers w.r.t. setup rate.]

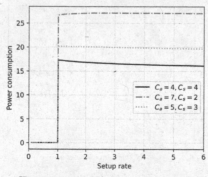

[Power consumption w.r.t. setup rate.]

Fig. 4. Effects of setup rate on performance measures

Example 2: Effects of the setup rate Fig. 4 shows how the mean number of customers and power consumption rate vary when α changes. In these numerical experiments, we set $\mu = 1$, $c = 5$, $R = 20$ and plot three curves in each case, corresponding to different values for C_a and C_s.

It is observed that when α is below 1 (the server setup times are very long), both considered measures are zero since customers do not find incentive to join the system. As α becomes larger, both measures increase. It is interesting to note that the mean number of customers in the system peaks at $\alpha = 2.3$, before slightly decreasing.

Example 3: Effects of the number of servers Fig. 5 illustrates how the number of servers affects the mean number of customers and the power consumption rate. In these numerical experiments, we set $\mu = 5$, $\alpha = 6$, $R = 20$ and plot three curves in each case, corresponding to different values for C_a and C_s.

[Mean number of customers w.r.t. Number of servers.]

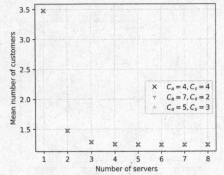

[Power consumption rate w.r.t.Number of servers.]

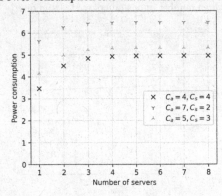

Fig. 5. Effects of the number of servers on performance measures

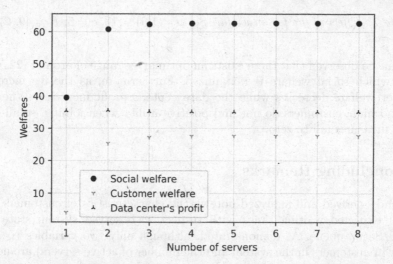

Fig. 6. Welfare w.r.t. Number of servers.

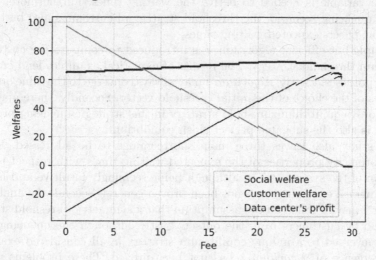

Fig. 7. Welfare w.r.t. fee.

It is seen that, as c increases, the mean number of customers in the system decreases, while the power consumption increases. These patterns are intuitive and similar to the outcomes in (Phung-Duc 2017).

Now, we examine the variation in welfare according to the number of servers. In the following example, $\mu = 5$, $\alpha = 6$, $R = 20$ and $C_a = C_s = 4$. We can observe from Fig. 6 that, while the profit of the data center is not so sensitive to the number of servers, both customer welfare and social welfare increase as more servers are added.

Example 4: Effect of the fee on welfare Set $\mu = 1$, $\alpha = 10$, $c = 5$, $R = 40$, $C_a = 8$ and $C_s = 2$.

It can be observed that there exists an optimal fee range of around $(22.5, 24)$, within which social welfare is maximized. Furthermore, as the fee increases, customer welfare decreases while the data center's profit increases. When the fee is too high, customers do not find positive utility when joining, so all these welfare measures fall to zero.

7 Concluding Remarks

This study derived and analyzed noteworthy characteristics of customers' conditional expected waiting times with respect to observed system state in a $M/M/c$/Setup queue. As demonstrated, although only two variables (i.e., the number of customers in the system and the number of active servers) are needed to model the system by a Markov chain and to derive performance measures, one more variable is needed to derive the waiting times. Furthermore, in the strategic queueing scenario, the threshold strategy of customers also becomes a factor that affects expected waiting times.

A simple but efficient algorithm was introduced to compute the customers' equilibrium threshold strategy. The results of numerical examples lend credence to two hypotheses. Firstly, *Algorithm 1* may always converge to the same outcome regardless of the choice of the initial threshold vector. Secondly, there may exist one and only equilibrium threshold strategy in this strategic queueing problem, and that is also the subgame perfect Nash equilibrium.

This study also leaves three main shortcomings to be addressed. Firstly, the monotonicity properties of the expected waiting times with respect to each dimension of the system states, although being seemingly intuitive and holding in all numerical examples, have not been proven general. Second, the analysis of customer behavior relied on the assumption that a symmetric threshold strategy is adopted by customers from the outset. Lastly, although the queueing game always converged to a unique equilibrium strategy in all considered examples, the convergence of *Algorithm* 1 has not been proved. These problems require further investigation.

Acknowledgements. The research of Hung Q. Nguyen was supported by JST SPRING, Grant Number JPMJSP2124. The research of Tuan Phung-Duc was supported in part by JSPS KAKENHI, Grant Number 21K11765 and by F-MIRAI: R&D Center for Frontiers of MIRAI in Policy and Technology, the University of Tsukuba and Toyota Motor Corporation collaborative R&D center.

References

Barroso, L.A., Hölzle, U.: The case for energy-proportional computing. Computer **40**(12), 33–37 (2007)

Gandhi, A., Doroudi, S., Harchol-Balter, M., Scheller-Wolf, A.: Exact analysis of the M/M/k/setup class of Markov chains via recursive renewal reward. Queueing Syst. **77**(2), 177–209 (2014). https://doi.org/10.1007/s11134-014-9409-7

Hassin, R., Haviv, M.: To Queue or not to Queue: Equilibrium Behavior in Queueing Systems, vol. 59. Springer Science & Business Media, Heidelberg (2003)

Hsieh, C.Y., Phung-Duc, T., Ren, Y., Chen, J.C.: Design and analysis of dynamic block-setup reservation algorithm for 5G network slicing. IEEE Trans. Mobile Comput. (2022)

Keilson, J., Servi, L.D.: A distributional form of Little's law. Oper. Res. Lett. **7**(5), 223–227 (1988)

Mitrani, I.: Service center trade-offs between customer impatience and power consumption. Perform. Eval. **68**(11), 1222–1231 (2011)

Mitrani, I.: Managing performance and power consumption in a server farm. Ann. Oper. Res. **202**(1), 121–134 (2013). https://doi.org/10.1007/s10479-011-0932-1

Naor, P.: The regulation of queue size by levying tolls. Econometrica **37**(1), 15–24 (1969)

Phung-Duc, T.: Exact solutions for M/M/c/setup queues. Telecommun. Syst. **64**(2), 309–324 (2017). https://doi.org/10.1007/s11235-016-0177-z

Phung-Duc, T., Kawanishi, K.: Delay performance of data-center queue with setup policy and abandonment. Ann. Oper. Res. **293**(1), 269–293 (2020). https://doi.org/10.1007/s10479-019-03268-1

Checkpointing Models for Tasks with Widely Different Processing Times

Paul Ezhilchelvan$^{(\boxtimes)}$ and Isi Mitrani

School of Computing, Newcastle University, Newcastle upon Tyne NE4 5TG, UK
{paul.ezhilchelvan,isi.mitrani}@ncl.ac.uk

Abstract. A server subject to random breakdowns and repairs offers services to incoming jobs whose lengths are highly variable. A checkpointing policy aiming to protect against possibly lengthy recovery periods is in operation. The problem of how to choose a checkpointing interval in order to optimize performance is addressed by analysing a general queueing model which includes breakdowns, repairs, back-ups and recoveries. Exact solutions are obtained under both Markovian and non-Markovian assumptions. Numerical experiments illustrate the conditions where checkpoints are useful and where they are not, and in the former case, quantify the achievable benefits.

Keywords: Breakdowns · Repairs · Recovery · M/G/1 queue · Embedded Markov chains · Server vacations · Checkpoint optimization

1 Introduction

Checkpointing is an important and useful crash-tolerant technique that involves storing process state during normal operation and restoring the recorded state to speed up recovery after a failure. It is cost-effective compared to hardware redundancy techniques, especially when the storage system for checkpointing data is reliable (Elnozahy et al. [11]). Not surprisingly, many commercial systems and public libraries, such as BlueGene/L (Adiga et al. [1]), IRIX OS (Tuthill et al. [25]) and Unix (Wang et al. [26]), have emerged to provide convenient APIs to facilitate its implementation.

The history of checkpointing goes back more than four decades, to the early days of transaction processing. Traditionally, the technique has involved keeping an 'audit trail' of transactions executed since the last checkpoint. Those transactions would be re-run in the event of a breakdown. The main question of interest would be how to choose the checkpoint frequency so as to minimize some appropriate cost function. In answering that question, the actual lengths of individual transactions were either ignored, or they were all assumed to have the same characteristics (e.g., exponentially distributed with the same mean).

We are interested in studying a checkpointing policy under a more realistic scenario where job processing times are random variables with a large coefficient of variation. That is, most of the jobs requiring service are short, but a few are very long. Exactly such a pattern of demand has been observed by monitoring a real-life cluster, see Chen et al. [5]. Under those conditions, the purpose of the policy would be to shorten the execution time of the long jobs by reducing the recovery period following a breakdown, without at the same time adding significantly to the processing of the short jobs. These considerations would govern the choice of the checkpoint interval.

The contribution of this paper is to analyse a queue served by a single unreliable server, operating a checkpoint policy in a mixed workload environment. That server is likely to be part of some distributed system. The incoming jobs are typically submitted by a web server. Processing a request may involve one or more database accesses whereby data is cached locally. All checkpoint back-ups take place on a database or possibly on the local disk.

A server breakdown may occur while (i) a job is being served, (ii) a checkpoint is being established, (iii) a recovery from a previous breakdown is in progress, or (iv) the queue is empty and the server is idle. Following a breakdown event, the server does nothing for a random interval which will be referred to as the 'repair period'. After that, in cases (i), (ii), or (iii), it performs a recovery operation consisting of going back to the last checkpoint if there is one, or to the beginning of the job if not, and redoing the work done since then. In case (iv), after being repaired the server takes the first waiting job, if any, and starts a new service.

Note that the repair period may not in fact involve an actual repair or reboot of the server. It may consist of disconnecting the primary server and replacing it with a secondary one that had been kept in reserve and in receipt of checkpoints from the primary (see Güler and Özkasap, [17], Oliveira et al. [22]). As far as the model is concerned, the exact nature of the operation is immaterial; of importance is only the distribution of the resulting inoperative interval.

The start of a job's execution plays the role of an initial checkpoint. Further checkpoints may be inserted at intervals as the run progresses. When the execution is completed, the job departs from the queue and a waiting job, if any, starts service. Thus, the checkpoint policy can be designed so as to leave short jobs largely unaffected, while reducing the run time of long jobs by shortening their recovery periods following a server breakdown.

Such a model has not, to our knowledge, been analyzed before. The objective of the analysis is to determine a performance measure such as the average response time or the average number of jobs present. This will enable the evaluation of the trade-offs between the costs incurred in backing-up the current process state, and the benefits derived from faster recovery operations. We start by solving the model under Markovian assumptions, but later generalize it to allow non-exponential distributions and also multiple servers.

1.1 Related Work

Models of checkpointing policies have been studied quite extensively over the years, under a variety of assumptions and application contexts. Research advances on checkpointing have also been 'checkpointed' by surveys at regular intervals. Worth mentioning are the surveys by Chandy [4], Nicola [21], Elnozahy et al. [11] and Marzouk and Jmaiel [19]. The Elnozahy et al. survey focusses exclusively on long-running computations, while Chandy mainly surveys checkpointing of streams of short transactions.

A large body of literature deals with long-running computations (sometimes referred to as 'infinite horizon'), motivated by scientific workloads which might typically take hours or days to complete. Those papers are not interested in performance metrics related to customers (e.g., average latency). The optimization criterion is the fraction of time that the server is doing useful work. Examples of such studies are Coffman and Gilbert [6], Liu et al. [18], Grassi et al. [16], Bruno and Coffman [3], Plank and Thomason [23], Subasi et al. [24] and Gelenbe et al. [15] (the last paper also aims to minimize the energy used).

More recently, Dimitriou [9] analysed a model where jobs finding a busy server are not queued, but retry after a random period. Such a policy would not be implemented in a transaction processing system because of its inefficient use of service capacity: the server may remain idle while jobs requiring service are present.

Models involving a queue of jobs have also been studied. Gelenbe [14] derived an expression for the optimal checkpoint interval. Baccelli [2] developed a numerical procedure for computing the average response time, while Dohi et al. [10] generalized the checkpoint policy by making it age-dependent. All those authors obtained their results by assuming that during operative periods the system behaves like an M/M/1 queue. Instead of an implicit checkpoint at the start of each job, an audit trail is maintained, keeping track of the jobs that would have to be re-run in the event of a breakdown. The consequence of that trail should be that results cannot be released to users, and jobs must be kept in the queue, until the next checkpoint is successfully established.

However, that is not what happens in the above models. Jobs are assumed to depart from the queue as soon as their service is completed. The recovery following a breakdown is simply a period during which jobs continue to arrive but none are served. The duration of that period is a linear function of the operative time elapsed since the last checkpoint.

In our model, departures upon service completion are justified by the fact that a breakdown only affects the job currently served, not the ones already completed.

The analysis of the Markovian model is described in Sect. 2, and its exact solution is presented in Sect. 3. The exact solutions when the back-up, checkpoint and repair intervals have general distributions, and the approximation for non-exponential intervals between breakdowns are described in Sect. 4. Several numerical experiments exploring the behaviour of the system for different parameter settings, under both Markovian and non-Markovian assumptions are shown in Sect. 5. These include an evaluation of the maximum achievable benefit of checkpointing.

2   Analysis Under Markovian Assumptions

The server goes through alternating periods of being operative and broken (or available and unavailable). These are distributed exponentially with means $1/\xi$ and $1/\eta$, respectively. Jobs arrive in a Poisson stream with rate λ. The required service times have a Hyperexponential distribution with K exponential phases, where phase k is entered with probability q_k and has an average of $1/\mu_k$ ($k = 1, 2, \ldots, K$). After completing the chosen exponential phase, the job departs.

A Hyperexponential distribution with K phases can be used to model jobs of K different types of the kind observed in [5]. Its coefficient of variation is always greater than or equal to 1, and can be arbitrarily large. For example, using just two Hyperexponential phases, with q_2 and μ_2 much smaller than q_1 and μ_1, respectively, one can model patterns of demand where most of the jobs are short and a few are very long.

While being served, a job sets up periodic checkpoints at random intervals. At the start of its service phase, or after a checkpoint has been established, a timer distributed exponentially with mean $1/\alpha$ is started. If that timer expires before the phase completes, a new checkpoint is attempted. The establishment of a checkpoint is not instantaneous but requires an exponentially distributed interval of time with mean $1/\beta$. That interval will be referred to as the 'back-up' operation.

Both the service intervals and the back-up operations may be interrupted by a server breakdown. Bearing in mind that the shortest of several exponentially distributed random variables is distributed exponentially with parameter equal to the sum of the parameters of the participating variables, we conclude that any service interval during phase k is distributed exponentially with parameter ν_k, given by

$$\nu_k = \mu_k + \alpha + \xi. \tag{1}$$

The end point of such an interval is either a service completion, with probability μ_k/ν_k, or a checkpointing attempt, with probability α/ν_k, or a server breakdown, with probability ξ/ν_k.

Similarly, any back-up operation in any phase is distributed exponentially with parameter σ, given by

$$\sigma = \beta + \xi. \tag{2}$$

Such an operation terminates with either the successful establishment of a checkpoint, with probability β/σ, or a server breakdown, with probability ξ/σ.

If the server breaks down during a service interval or during a following back-up operation, the work performed since the last checkpoint, or in the absence of a checkpoint since the beginning of the phase, must be repeated when the server is repaired. This is referred to as the 'recovery' operation. According to the above observations, the recovery operations in phase k are distributed exponentially with parameter ν_k. Of course, the server may break down again during a recovery, in which case another recovery (depending on the phase) is started after the repair.

We assume that each recovery operation is a new sample from the appropriate distribution. That assumption is motivated by the fact that different runs of the same task never take exactly the same time, particularly in a multi-core environment.

Note that the action taken after the server is repaired following a breakdown depends on whether the breakdown occurred while the server was idle, or whether it occurred while the server was active (i.e., serving a job, backing-up or recovering). In the former case there is no need for a recovery: either the server is again idle, or a job has arrived in the meantime and a new service begins. In the latter case, a new recovery starts, whose duration depends on the phase that was in progress when the breakdown occurred.

Denote by T the random variable representing the total period between the start of a job's service and its completion. That period includes service intervals and back-up operations, as well as repair times and recovery operations following any breakdowns. The interval T will be referred to as the 'effective service time'. We shall need the first and second moments of that interval. In particular, the necessary and sufficient condition for stability of the system is that the offered load generated by the effective service times of the incoming jobs should be less than 1:

$$\lambda E(T) < 1. \tag{3}$$

Let X_k be the time a job takes to complete phase k. The moments of the effective service time are simply expressed in terms of the moments of X_k:

$$E(T) = \sum_{k=1}^{K} q_k E(X_k), \tag{4}$$

and

$$E(T^2) = \sum_{k=1}^{K} q_k E(X_k^2). \tag{5}$$

3   Exact Solution

To determine the moments of the effective service time, we shall introduce two sets of auxiliary random variables. Define Y_k as the interval between attempting to establish a checkpoint during phase k, and resuming phase k service. That interval may include repairs and recovery operations resulting from breakdowns during the back-up operation. Also define V_k as the interval between a breakdown in phase k and resuming phase k service. It includes the repair and the recovery operation, plus any additional repairs and recoveries caused by further breakdowns. Denote by $y_k(s)$ and $v_k(s)$ the Laplace transforms of the Y_k and V_k probability density functions, respectively.

The Laplace transform of an exponential p.d.f. with some parameter, γ, is equal to $\gamma/(\gamma+s)$. Also, the transform of a sum of independent random variables is equal to the product of their transforms. Hence, we can write the following equation for $v_k(s)$:

$$v_k(s) = \frac{\eta}{\eta + s} \frac{\nu_k + \xi}{\nu_k + \xi + s} \left[\frac{\nu_k}{\nu_k + \xi} + \frac{\xi}{\nu_k + \xi} v_k(s) \right], \tag{6}$$

where ν_k is given by (1). The first term in the square brackets is the probability that the recovery operation completes without interruption; the second term contains the probability that another breakdown occurs and a new random V_k is started.

The first and second moments of V_k are obtained from $E(V_k) = -v_k'(0)$ and $E(V_k^2) = v_k''(0)$. Differentiating (6) twice at $s = 0$ yields, after some algebra,

$$E(V_k) = \frac{\nu_k + \xi + \eta}{\nu_k \eta}, \tag{7}$$

and

$$E(V_k^2) = 2 \left[E(V_k)^2 - \frac{1}{\nu_k \eta} \right]. \tag{8}$$

Since the interval Y_k terminates either as a successful back-up, or is interrupted by a breakdown and is followed by a recovery operation V_k, we can express $y_k(s)$ in terms of $v_k(s)$:

$$y_k(s) = \frac{\sigma}{\sigma + s} \left[\frac{\beta}{\sigma} + \frac{\xi}{\sigma} v_k(s) \right], \tag{9}$$

where σ is given by (2). The first and second moments of Y_k are given by

$$E(Y_k) = \frac{1}{\sigma} [1 + \xi E(V_k)], \tag{10}$$

and

$$E(Y_k^2) = \frac{2}{\sigma^2} [1 + \xi E(V_k)] + \frac{\xi}{\sigma} E(V_k^2). \tag{11}$$

Now remember that the phase k execution, X_k, consists of a service interval which either terminates uninterrupted, or is interrupted by a back-up operation, Y_k, and later resumed, or is interrupted by a breakdown interval, V_k, and later resumed. This leads to the following equation for the Laplace transform of X_k, $x_k(s)$.

$$x_k(s) = \frac{\nu_k}{\nu_k + s} \left[\frac{\mu_k}{\nu_k} + \frac{\alpha}{\nu_k} y_k(s) x_k(s) + \frac{\xi}{\nu_k} v_k(s) x_k(s) \right]. \tag{12}$$

Differentiating twice at $s = 0$ and substituting the moments of Y_k and V_k already derived, we obtain the first and second moments of X_k:

$$E(X_k) = \frac{1}{\mu_k} [1 + \alpha E(Y_k) + \xi E(V_k)], \tag{13}$$

and

$$E(X_k^2) = 2E(X_k)^2 + \frac{1}{\mu_k} [1 + \alpha E(Y_k^2) + \xi E(V_k^2)]. \tag{14}$$

Using expressions (10) and (7), the average execution time of phase k can be rewritten as

$$E(X_k) = \frac{1}{\mu_k} \frac{\xi + \eta}{\eta} \frac{\alpha + \sigma}{\sigma} \frac{\nu_k + \xi}{\nu_k}, \tag{15}$$

The ergodicity condition (3) can now be stated explicitly:

$$\sum_{k=1}^{K} q_k \rho_k \frac{\nu_k + \xi}{\nu_k} < \frac{\eta}{\xi + \eta} \frac{\sigma}{\alpha + \sigma}, \tag{16}$$

where $\rho_k = \lambda / \mu_k$. Note that when $\alpha = 0$ and $\xi = 0$, i.e. when there are no checkpoints and no breakdowns, this reduces to the usual stability condition for the M/G/1 queue: the offered load must be less than 1.

The exact solution for our model cannot be obtained by treating it as a simple M/G/1 queue. This is because of the possibility that the server may break down while the queue is empty. If that happens, a job may arrive into the system, find an empty queue, yet be unable to start service immediately. This situation is similar to what is known in the literature as a 'server of walking type', or 'server with vacations', where a server encountering an empty queue goes away for a random period (e.g., see [12]). In those studies, vacations are always taken, and are independent of the arrival process. Our model is different, in that the server only takes a vacation (a repair period) if a breakdown occurs while the queue is empty.

For an exact analysis, we shall consider the number of jobs present at (just after) consecutive departure instants. This is a discrete time Markov chain. The following notation will be used:

π_i is the steady-state probability that there are i jobs left in the system after a departure ($i = 0, 1, \ldots$). This is also the probability that an incoming job would see i jobs present. Hence, by the PASTA property, it is also the probability that a random observer would see i jobs present.

a_i is the probability that i jobs arrive during an effective service time, T;

b_i is the probability that i jobs arrive during a repair period;

We shall also introduce the generating functions

$$\pi(z) = \sum_{i=0}^{\infty} \pi_i z^i; \quad a(z) = \sum_{i=0}^{\infty} a_i z^i; \quad b(z) = \sum_{i=0}^{\infty} b_i z^i. \tag{17}$$

If $i > 0$ and k jobs arrive during the ensuing service time ($k = 0, 1, \ldots$), then the next state will be $i + k - 1$. If $i = 0$, then the next state will also depend on whether the idle period is interrupted by a breakdown or not. Transforming the relevant probabilities and balance equations into generating functions, we obtain, after some work, the following equation for $\pi(z)$ in terms of $a(z)$ and $b(z)$:

$$\pi(z) = \frac{(1 - \rho)a(z)(z - 1)}{z - a(z)} \left[\frac{\eta}{\xi + \eta} + \frac{\xi}{\xi + \eta} \frac{\eta}{\lambda + \eta} \frac{zb(z) - 1}{z - 1} \right]. \tag{18}$$

This representation is significant, because it expresses $\pi(z)$ as a product of two generating functions. The first fraction in the right-hand side of (18) is the well-known generating function of the steady-state distribution of the M/G/1 queue (e.g., see [7]). The term in the square brackets is the generating function of a random variable which we shall call U; it is related to the number of arrivals during a repair period. That product form implies the following result:

Lemma. *The number of jobs present in our system in the steady state is distributed as the sum of the number of jobs in the corresponding M/G/1 queue (arrival rate λ and service time T), plus the random variable U.*

Similar results can be found in the literature related to models of servers with vacations.

In particular, denoting by L and $L_{M/G/1}$ the steady-state average numbers of jobs in our system and in the corresponding M/G/1 queue respectively, we can write

$$L = L_{M/G/1} + E(U). \tag{19}$$

When the repair periods are distributed exponentially, the distribution b_i is geometric with parameter $\lambda/(\lambda + \eta)$. The generating function $b(z)$ is

$$b(z) = \frac{\eta}{\lambda(1-z) + \eta}. \tag{20}$$

Substituting this into the generating function of U and differentiating at $z = 1$, we obtain

$$E(U) = \frac{\lambda\xi}{\eta(\xi + \eta)}. \tag{21}$$

This is quite an intuitive expression: it represents the average number of arrivals during a repair period (λ/η), multiplied by the fraction of time during which the server is broken.

Equation (19), together with Pollaczek-Khinchin's result (e.g., see [20]), now provides the average L in terms of quantities that have already been derived:

$$L = \lambda E(T) + \frac{\lambda^2 E(T^2)}{2(1 - \lambda E(T))} + \frac{\lambda\xi}{\eta(\xi + \eta)}, \tag{22}$$

where $E(T)$ and $E(T^2)$ are given by (4) and (5).

This completes the exact solution of the model under Markovian assumptions. It is worth mentioning that other phase-type distributions of the required service times can be handled by the same methods. For example, in some applications it may be more convenient to assume a Coxian distribution with K exponential phases, where phase k has an average of $1/\mu_k$ and is followed by phase $k + 1$ with probability q_k. After completing phase K, the job departs. A three-phase Coxian distribution is illustrated in Fig. 1.

The class of Coxian distributions is, for most purposes, general (see [8]), and can also model mixed job populations.

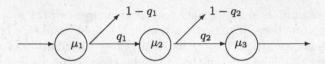

Fig. 1. A Coxian distribution with three phases

Let r_k be the probability that a job reaches phase k during the course of its service:

$$r_1 = 1; \quad r_k = \prod_{i=1}^{k-1} q_i; \quad k = 2, 3, \ldots, K. \tag{23}$$

Then the moments of the effective service time, T, are obtained from

$$E(T) = \sum_{k=1}^{K} r_k E(X_k), \tag{24}$$

and

$$E(T^2) = \sum_{k=1}^{K} r_k E(X_k^2) + 2 \sum_{k=2}^{K} r_k \sum_{i=1}^{k} E(X_i) E(X_k). \tag{25}$$

The analysis presented above would apply, provided that each new phase begins with a checkpoint.

4   Generalizations

It is possible, and desirable, to relax a number of the assumptions that have been made so far. In the majority of cases this can be done while retaining the exact nature of the solution. However, one of the generalizations appears to be intractable and requires an approximation.

4.1   General Repair, Backup and Checkpoint Intervals

Suppose that the repair period, R, has a general distribution, with Laplace transform $\eta(s)$. The exact analysis can proceed largely as before, subject to two important changes. First, Eq. (6) becomes

$$v_k(s) = \eta(s) \frac{\nu_k + \xi}{\nu_k + \xi + s} \left[\frac{\nu_k}{\nu_k + \xi} + \frac{\xi}{\nu_k + \xi} v_k(s) \right]. \tag{26}$$

Given the first two moments of the repair time, $E(R)$ and $E(R^2)$, this expression enables us to evaluate $E(V_k)$ and $E(V_k^2)$, and hence proceed to determine $E(X_k)$ and $E(X_k^2)$.

The second change is that the argument leading to (22) should now be generalized to include random vacations with general distribution. The analysis of the previous section can be modified to produce a generalized form of expression (18), involving the Laplace Transform of the Random Modification of the repair period (i.e. the remaining duration of the repair period, as seen by a random observer). The resulting quantity $E(U)$, which appears in (19) and (21) now contains the first and second moments of the repair time:

$$E(U) = \lambda \frac{E(R^2)}{2} \frac{\xi}{1 + \xi E(R)}. \tag{27}$$

Similar arguments can be employed when the back-up times and the checkpoint intervals have general distributions with Laplace transforms $b(s)$ and $a(s)$, respectively. They lead to a generalized form of Eq. (12), from which the moments $E(X_k)$ and $E(X_k^2)$ are determined by differentiation at $s = 0$. The details of those developments are omitted for lack of space.

4.2   Approximation for General Breakdown Intervals

Relaxing the Markovian character of the breakdown process, while still computing an exact solution, appears to be considerably more difficult. We are therefore proposing a simple approximation that can be readily justified in a practical situation. The argument is based on the fact that breakdowns are rare events.

Assume that the operative periods of the server, i.e. the intervals between completing a repair and the next breakdown, are i.i.d. random variables with an m-phase Coxian distribution. That is sufficiently general for practical purposes. By choosing the number of phases and their parameters appropriately, one can approximate most commonly used distributions, such as Erlang, Weibull, Hyperexponential, Lognormal, Normal, etc.

Denote by ξ_i the parameter of phase i $(i = 1, 2, \ldots, m)$, by $\bar{q}_i$ the probability of moving from phase $i - 1$ to phase i, and by $\bar{r}_i = \bar{q}_1 \bar{q}_2 \cdots \bar{q}_{i-1}$ the probability of reaching phase i.

It is reasonable to assume that the parameters ξ_i are small compared to the other parameters (i.e. the average lengths of all phases are large). In that case, the queueing process can be assumed to be stable and reach steady state during each phase. The approximate solution of the generalized model would then consist of the following steps.

1. Compute the steady-state probabilities, γ_i, that an operative server is in phase i of its breakdown interval. These probabilities satisfy the balance equations

$$\gamma_i \xi_i = \gamma_{i-1} \xi_{i-1} \bar{q}_{i-1}; \quad i > 1. \tag{28}$$

They can therefore be expressed, after normalization, as

$$\gamma_i = \frac{\xi_1 \bar{r}_i}{\xi_i} \left[\sum_{j=1}^m \frac{\xi_1 \bar{r}_j}{\xi_j} \right]^{-1}; \quad i = 1, 2, \ldots, m. \tag{29}$$

2. Apply the existing exact solution to phase i, using $\xi = (1 - \bar{q}_i)\xi_i$ as the breakdown rate, and compute the average number of jobs present, L_i.
3. Compute the overall average number of jobs present, L, as a weighted average over all phases:

$$L = \sum_{i=1}^{m} \gamma_i L_i. \tag{30}$$

Remark. The above approximation, known as 'decomposition', has been used in a variety of other contexts. A quantitative evaluation of its accuracy for our system may be provided by simulations. However, as the breakdowns are rare events, that would be a non-trivial exercise, possibly requiring special techniques. Such an undertaking is outside the scope of the present work.

5   Numerical Results

The exact solutions of Sects. 3 and 4 were applied to several example systems, with the aim of examining the trade-offs between the costs and benefits of checkpointing. In order to reduce the parameter space to be explored, some of the parameters are kept fixed. The required service times are assumed to have a two-phase Hyperexponential distribution with quite a large coefficient of variation: $1/\mu_1 = 40$; $1/\mu_2 = 400$; $q = 0.2$. In other words, the average requirement of 80% of the incoming jobs is 40, and for the other 20% it is 400. These parameters are chosen to conform, as far as it was possible to extract average values from the reported statistics, with the data collected in [5]. The average repair period is assumed to be relatively short, $1/\eta = 15$. The arrival rate λ, the checkpointing rate, α, the average back-up interval. $1/\beta$ and the breakdown rate, ξ, are varied. The performance measure in all cases is the average number of jobs present, L. If we wished to evaluate the average response tme, W, we would use Little's result to compute $W = L/\lambda$.

In the first example, the arrival rate is set to $\lambda = 0.0065$, and L is plotted against α, for three different values of the breakdown rate: $\xi = 10^{-4.5}$, $\xi = 10^{-4}$ and $\xi = 10^{-3.5}$. These rates are comparable to the ones reported in Garraghan et al. [13].

With these parameters, the system load, as measured by $\lambda E(T)$, is on the order of 75% or higher.

The results are shown in Fig. 2. All three plots exhibit a steep initial decline in occupancy, followed by a slow increase. As the breakdown rate increases, the optimal α also increases slightly; it is about 0.1 when $\xi = 10^{-4.5}$, between 0.1 and 0.2 when $\xi = 10^{-4}$ and close to 0.3 when $\xi = 10^{-3.5}$. Such an increase could have been expected, since a more likely breakdown during service makes the backing-up of the current state more advantageous. The low gradient of the plateau following the optimal point is explained by the low cost of the back-up operation. Of course, if α carries on increasing, so that the 'non-productive' back-up operations push the queue closer to saturation, the increase in occupancy would accelerate without bound.

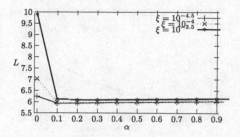

Fig. 2. $\lambda = 0.0065$, different ξ

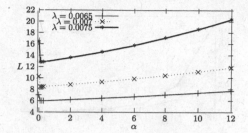

Fig. 3. $\xi = 10^{-4}$, different λ.

This phenomenon is illustrated in the second experiment, where the breakdown rate is kept fixed at $\xi = 10^{-4}$, while the arrival rate takes three different values: $\lambda = 0.0065$, $\lambda = 0.007$ and $\lambda = 0.0075$. The results are shown in Fig. 3. Again, the plots of L against α display a steeply decreasing portion, followed by an increasing one.

For all three values of λ, the optimal checkpointing rate is between $\alpha = 0.1$ and $\alpha = 0.2$. However, having extended the range of α considerably beyond the optimum, we observe the non-linear increase in L caused by the back-up operations. This is particularly noticeable in the case of $\lambda = 0.0075$, where the system was quite heavily loaded to start with.

It is intuitively clear that the more reliable the server, the less frequent need be the checkpoints. On the other hand, the less time it takes to perform a back-up operation, the more checkpoints can be afforded. In order to quantify these observations, we have evaluated the optimal checkpointing rate, α^*, as a function of the average interval between breakdowns, $1/\xi$. This was done for three different values of the back-up rate, $\beta = 100$, $\beta = 200$ and $\beta = 400$. The results are presented in Fig. 4, where $1/\xi$ is scaled exponentially. At each point, the optimal α^* was found by carrying out a search.

As expected, the optimal checkpointing rate eventually becomes zero when the server becomes sufficiently reliable. However, that point is not reached quickly: when $\beta = 100$ or $\beta = 200$, the mean time between failures needs to be about 10^7 before checkpoints become counterproductive. That interval increases to 10^8 for $\beta = 400$. At the other end of the range, when breakdowns are relatively frequent, the optimal checkpointing rate is higher and increases with the speed of the back-up operations.

In order to examine the effect that a change of distributions has on performance, we have repeated the experiment of Fig. 4, under the assumptions that the repair, back-up and checkpoint intervals are constant, keeping the means, $E(R)$, $E(B)$ and $E(A)$, as before. Clearly, such a change would reduce the variance of the effective service time and hence would reduce the average number L. However, it is not obvious whether the optimal value of the checkpoint frequency, $\alpha = 1/E(A)$, is affected and if so, to what extent.

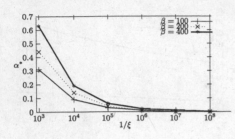

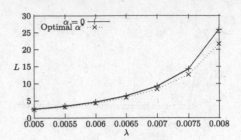

Fig. 4. Optimal α **Fig. 5.** Benefits of checkpointing

We observed a similar behaviour to the one exhibited in Fig. 4: as the server becomes more reliable, the optimal checkpoint frequency decreases and eventually becomes 0 (i.e. checkpoints are no longer needed).

The notable difference in these observations is that the values of α^* were in all cases significantly lower than before. That outcome can be explained intuitively by arguing that a deterministic checkpointing policy is more effective than a random one with the same averages. Hence, one can achieve the desired result with fewer checkpoints.

Because of the similarity of behaviours, it was not deemed necessary to include another version of the figure.

The final experiment aims to quantify the gains achievable by a checkpointing policy. This is done by comparing the average number of jobs present, L, when no checkpoints are used (i.e. $\alpha = 0$ or $a = \infty$), with the number present when the policy uses the optimal checkpoint frequency α^*. The latter value is determined by a search. The other parameters are as before, with $\xi = 10^{-4}$. The repair, back-up and checkpoint intervals are constant.

In Fig. 5, the unoptimized and optimized values of L are plotted against the arrival rate, which varies from $\lambda = 0.005$ to $\lambda = 0.008$. On that range, the offered load, λT, varies from about 55% to about 90%.

As might have been expected, when the system is lightly loaded, the gains of checkpointing are slight. Jobs affected by breakdowns do not tend to delay other jobs at light loads, because the queue is short. Hence, any savings in their effective service times show limited benefits. As the load increases, the queue gets longer and the savings are noticed by more waiting jobs. At the 90% load, the difference between no checkpointing and optimal checkpointing is about 25%.

We carried out the same comparison under the assumption that the A, B and R random variables are distributed exponentially. The results were very similar to the ones presented in Fig. 5 and are therefore omitted.

6   Conclusions

We have examined the effects of checkpointing on performance by analysing a rather general queueing model involving breakdowns, repairs and back-up operations. A major objective of the study was to handle a job population with a large variability of required service times. Exact solutions were obtained under both Markovian and non-Markovian assumptions. These were used in order to determine the optimal checkpoint frequency for different parameter settings, and to quantify the benefits of checkpointing.

One of the proposed generalizations, to non-exponential intervals between breakdowns, involves an approximate solution. As indicated in the Remark at the end of Subsect. 4.2, assessing the accuracy of that approximations is outside the scope of the present paper but would be a worthy topic for future research.

Another interesting generalization that would be worth studying is to divide the mixed job population into several classes, with a separate queue for each class. A non-preemptive priority scheduling policy could be in operation among the classes. For example, short jobs might be given higher priority than long ones. It may be possible to generalize existing results on multi-class M/G/1 queues to such a system. To obtain an exact solution, it would be necessary to analyse the effect that breakdowns during idle periods have on the behaviour of queues. That too would be a worthy topic for future research.

References

1. Adiga, N.R., et al.: An overview of the BlueGene/L supercomputer. In: Proceedings of the ACM/IEEE Conference on Supercomputing, p. 60 (2002). https://doi.org/10.1109/SC.2002.10017
2. Baccelli, F.: Analysis of a service facility with periodic checkpointing. Acta Inform. **15**, 67–81 (1981)
3. Bruno, J.L., Coffman, E.G.: Optimal fault-tolerant computing on multi-processor systems. Acta Inform. **34**, 881–904 (1997)
4. Chandy, K.M.: A survey of analytic models of rollback and recovery strategies. Computer **8**(5), 40–47 (1975)
5. Chen, Y., Ganapathi, A.S., Griffith, R., Katz, R.H.: Analysis and lessons from a publicly available google cluster trace. Technical report No. UCB/EECS-2010-95 (2010)
6. Coffman, E.G., Gilbert, E.N.: Optimal strategies for scheduling checkpoints and preventive maintenance. IEEE Trans. Reliabil. **39**(1), 9–18 (1990)
7. Cohen, J.W.: The Single Server Queue. North-Holland, Amsterdam (1969)
8. Cox, D.R.: A use of complex probabilities in the theory of stochastic processes. Math. Proc. Camb. Philos. Soc. **51**(2), 313–319 (1955)
9. Dimitriou, I.: A retrial queue for modeling fault-tolerant systems with checkpointing and rollback recovery. Comput. Ind. Eng. **79**, 156–167 (2015)
10. Dohi, T., Kaio, N., Trivedi, K.S.: Availability models with age-dependent checkpointing. In: 21st IEEE Symposium on Reliable Distributed Systems, pp. 130–139 (2002)
11. Elnozahy, E.N., Alvisi, L., Wang, Y., Johnson, D.B.: A survey of rollback-recovery protocols in message-passing systems. ACM Comput. Surv. **34**(3), 375–408 (2002)
12. Fuhrmann, S.W.: A note on the M/G/1 queue with server vacations. Oper. Res. **32**(6), 1368–1373 (1984)
13. Garraghan, P., Townend, P., Xu, J.: An empirical failure-analysis of a large-scale cloud computing environment. In: 15th International Symposium on High-Assurance Systems Engineering, pp. 113–120 (2014)

14. Gelenbe, E.: On the optimum checkpoint interval. J. ACM **26**(2), 259–270 (1979)
15. Gelenbe, E., Boryszko, P., Siavvas, M., Domanska, J.: Optimum checkpoints for time and energy. In: 28th IEEE Symposium on Modeling, Analysis, and Simulation of Computer and Telecommunication Systems (MASCOTS), pp. 1–8 (2020)
16. Grassi, V., Donatiello, L., Tucci, S.: On the optimal checkpointing of critical tasks and transaction-oriented systems. IEEE Trans. Softw. Eng. **18**(1), 72–77 (1992)
17. Güler, B., Özkasap, Ö.: Efficient checkpointing mechanisms for primary-backup replication on the cloud. Concurr. Comput. Pract. Exp. **30**, 21 (2018)
18. Liu, Y., Nassar, R., Leangsuksun, C., Naksinehaboon, N., Paun, M., Scott, S.L.: An optimal checkpoint/restart model for a large scale high performance computing system. In: IEEE Symposium on Parallel and Distributed Processing, pp. 1–9 (2008)
19. Marzouk, S., Jmaiel, M.: A survey on software checkpointing and mobility techniques in distributed systems. Concurr. Comput. Pract. Exp. **23**(11), 1196–1212 (2011)
20. Mitrani, I.: Probabilistic Modelling. Cambridge University Press, Cambridge (1998)
21. Nicola, V.F.: Checkpointing and the modelling of program execution time. In: Lyu, M.R. (ed.) Software Fault Tolerance, pp. 167–188. Wiley (1995)
22. Oliveira, R., Pereira, J., Schiper, A.: Primary-backup replication: from a time-free protocol to a time-based implementation. In: Proceedings of the 20th IEEE Symposium on Reliable Distributed Systems, pp. 14–23 (2001)
23. Plank, J.S., Thomason, M.G.: Processor allocation and checkpoint interval selection in cluster computing systems. J. Parallel Distrib. Comput. **61**(11), 1570–1590 (2001)
24. Subasi, O., Kestor, G., Krishnamoorthy, S.: Toward a general theory of optimal checkpoint placement. In: IEEE Conference on Cluster Computing (CLUSTER), pp. 464–474 (2017)
25. Tuthill, B., Johnson, K., Schultz, T.: Irix checkpoint and restart operation guide. Document of Silicon Graphics Inc. (1999)
26. Wang, Y.-M., Huang, Y., Vo, K.-Ph., Chung, P.-Y., Kintala, C.: Checkpointing and its applications. In: 25th International Symposium on Fault-Tolerant Computing. Digest of Papers, pp. 22–31 (1995)

Machine Learning

Analysis of Reinforcement Learning for Determining Task Replication in Workflows

Andrew Stephen McGough[✉][iD] and Matthew Forshaw[iD]

School of Computing, Newcastle University, Newcastle upon Tyne, UK
{stephen.mcgough,matthew.forshaw}@ncl.ac.uk

Abstract. Executing workflows on volunteer computing resources where individual tasks may be forced to relinquish their resource for the resource's primary use leads to unpredictability and often significantly increases execution time. Task replication is one approach that can ameliorate this challenge. This comes at the expense of a potentially significant increase in system load and energy consumption. We propose the use of Reinforcement Learning (RL) such that a system may 'learn' the 'best' number of replicas to run to increase the number of workflows which complete promptly whilst minimising the additional workload on the system when replicas are not beneficial. We show, through simulation, that we can save 34% of the energy consumption using RL compared to a fixed number of replicas with only a 4% decrease in workflows achieving a pre-defined overhead bound.

Keywords: Performance · Reinforcement learning · Scheduling

1 Introduction

Workflows comprising several independent (computational) tasks under a strict ordering (see Fig. 1 for an example) have become one of the pillars of computational research and industrial development. Users wish to enact these as quickly as possible – within a factor of the critical path execution time – referred to as the contingency. However, many organisations lack dedicated resources, or the operational budget for Cloud enactment. Here these users are required to 'make do' with computational resources which are not dedicated to this purpose – often referred to as volunteer computing. One of the common ways of providing this is through High Throughput Computing (HTC), which exploits the idle time available on computing resources provisioned for other purposes – examples of such systems include HTCondor [34] and BOINC [1].

Although HTC systems have marginal capital expense, they have the distinct disadvantage that resources are often retracted without warning for their primary use. This will lead to the termination of those tasks which are currently being executed on that resource, and potentially significant delays in workflow execution. Approaches to ameliorate this include suspension of tasks until the resource

K. Gilly and N. Thomas (Eds.): EPEW 2022, LNCS 13659, pp. 117–132, 2023.
https://doi.org/10.1007/978-3-031-25049-1_8

is available again [32], relaunching the task on a new resource to run from the beginning [24] or checkpointing and migration [11,31]. These approaches ensure eventual task completion, but impact the overall execution time of the task [29] – which can be compounded when enacting a workflow.

Users will, in general, expect their workflows to finish in a "reasonable" time, relative to execution time for the workflow. Given that the tasks within the workflow will often have to relinquish the resource back to the primary user there can be considerable variance to the workflow execution time (see Sect. 2). Leading to users believing that their workflow has crashed or the system is unfit for purpose. However, this could just be a consequence of 'bad luck' for the particular workflow enactment. Ideally a workflows execution time should be relative to the length of the workflow's critical path (defined as the shortest time from the start to the end of the workflow [16]) where a subset of the workflow tasks form the critical path. We formally state the users desire here that a given workflow will complete within time $(1 + p)CR(W)$, where the delay proportion $p \geq 0$ and $CR(W)$ is the time to execute the critical path for workflow W.

Although a task may not initially form part of the critical path if it is forced to relinquish a resource the delay to its completion may make it part of a newly revised critical path. As such, the workflow enactment needs either to reduce the chance of the task from becoming so delayed or compensate for such a delay.

Task replication – running multiple copies of the same task – can ameliorate the impact of such delays [27]. However, this can quickly impact other work on the cluster leading to all workflows taking longer to complete and higher overall energy consumption – as all but one are wasted work. Determining how many replicas to have is a difficult problem and depends on the resources primary usage pattern, how critical the task is to timely completion and other workloads on the system. We propose using Reinforcement Learning (RL) [33] to 'learn' the number of replicas to deploy based on the time of day and how critical a tasks is to the workflow finishing within its contingency requirements.

Section 2 discusses the problem space and analyse the impact of running workflows in non-dedicated environments. Related work is evaluated in Sect. 3. Our approach is presented in Sect. 4. Section 5 presents our simulation model before presenting results in Sect. 6. We present our conclusions in Sect. 7.

2   Problem Space Analysis

We define, without loss of generality, a workflow as a Directed Acyclic Graph (DAG) comprising independent tasks (represented as nodes) with an associated ordering on their execution (represented by directed links between nodes). Tasks have an associated execution time, including time for setup, data ingress and egress, whilst links have associated time required for scheduling. A DAG may have more than one start/end node. Without compromising our model, we assume that such DAGs have an extra 'task' at the start (end), providing a single start (end) node. These tasks have zero execution time; however, the link execution time may be non-zero – to cater for workflow setup or termination.

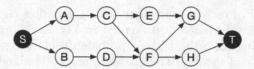

Fig. 1. A workflow. Nodes are tasks to run and links represent dependencies.

Figure 1 illustrates a simple DAG. The normal tasks (A–H) have links to indicate their dependencies (e.g. task F can only execute once both tasks C and D have completed). Extra tasks are added to give a single start (S) and termination (T).

We analyse the impact on a workflow when running it on volunteer resources such as a University HTC system. We extend HTC-Sim [9] for workflows [26]. As the original trace-log does not contain workflows we replace 10% of the original tasks with synthetic workflows (a simple workflow – that of Fig. 1 with each task taking 32 min). For each workflow which completed we compute the value of $p(W)$, the proportion of excess time to execute the workflow, as:

$$p(W) = \frac{(c(W) - s(W))}{CR(W)} - 1 \tag{1}$$

where $c(W)$, $s(W)$ are the finish and submission times of tasks T, S from W and $CR(W)$ is the execution time for the critical path of W.

Simple statistical analysis (mean 0.4467, standard deviation 0.8217) of the 40,054 workflows suggests little impact from running in a volunteer environment. However, the Cumulative Density function (Fig. 2a) illustrates the long-tail on these results. Very few workflows complete with a value of p(W) less than 0.00004. With a significant ($\sim$30%) of workflows having $p(W) \sim 0.000043$. On the other hand, 10% of the workflows take double or more than their critical path time to run ($p(W) \geq 1$). These would seem to match those situations where the system was under exceptional load causing workflow tasks to be repeatedly forced to relinquish resources. Our worst case was when $p(W) = 17.1113$.

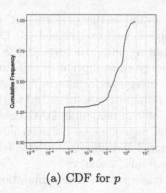

(a) CDF for p

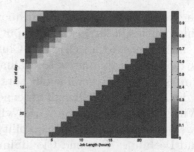

(b) Probability of task completing based on length and time of submission

Fig. 2. CDF and probability of task completion

The probability of each individual task in a workflow completing given its length and time of submission is presented in Fig. 2b. Short tasks submitted early in the morning have the highest probability of success with reasonable chances across most of the day. The discontinuity at 3am and the reduction to zero probability for longer jobs later in the day is a consequence of the nightly reboot at 3am. For this system we predict that replicas are likely to be beneficial through the day, but less so in the early morning or before the 3am reboot. In this work we propose an adaptive system which 'learns' system-specific behaviours.

3   Related Work

3.1   Energy-Aware Workflow Scheduling

The challenges of scheduling workflows to computational resources has received extensive treatment in the literature. A detailed exposition of previous efforts is presented by Yu *et al.* [36]. Existing approaches predominantly focus on workflow scheduling performance, with comparatively few considering the energy consumption as a primary optimisation goal. Durillo *et al.* [8] adapt the well-established HEFT [35] heuristic for workflow scheduling, and demonstrate the potential for significant energy conservation, for a small performance penalty.

Several approaches explore enacting workflows in the Cloud. Zhu *et al.* [37] propose *pSciMapper* for energy-aware consolidation of workflow tasks in virtualised environments. Li *et al.* [21] adopt a particle swarm optimisation (PSO) approach to workflow scheduling in the Cloud, with consideration for financial cost and security. Calheiros *et al.* [6] focus on task replication to overcome issues introduced by performance variation of public Cloud resources.

Our approach is distinct from those presented. Firstly, we consider workflow enactments to non-dedicated resources shared with their primary users. Secondly, we adopt a RL based approach to learn characteristics of the environment.

3.2   Simulations with Support for Workflows

The benefits of using simulation frameworks to evaluate the performance of scheduling policies in large-scale computing systems is well understood [10,12]. Several simulators exist for Grid and Cluster computing infrastructures [2,4,20], but lack modelling capabilities to evaluate the energy consumption of the simulated infrastructure. Several simulators evaluate energy-performance [5,19,22, 30], but do not allow modelling of tasks needing to relinquish resources and workflows. With the exception of SimGrid, which is capable of modelling DAG-based workflows, these tools lack the capability to model dependencies between jobs. Chen and Deelman [7] extend CloudSim [5] to model Workflows, and evaluate the performance of several common workflow scheduling heuristics.

Previously we extended HTC-Sim [13], a Java-based trace-driven simulation, to support workflow execution [26]. Here we explore the energy and performance impact of workflow enactment schemes within a multi-use cluster environment.

4   Reinforcement Replication

We present our replication approach within a volunteer HTC environment, however, it is equally applicable to other environments exhibiting intermittent faults.

4.1   Approach Overview

We augment an event based model [14] for workload enactment to support task replication, in which every time the status of the workflow changes (e.g. workflow starts, task completes) an event is triggered and the appropriate actions are performed. We assume here that a user has defined ϕ[1] their contingency proportion for workflow execution. Our adaptations to these workflow event handlers are:

– **Workflow starts:** The desired deadline d for the workflow is computed:

$$d = (\phi + 1)CR(W) + s(W). \tag{2}$$

Next, for each task t within the workflow which can be run – effectively those tasks linked to task 'S' – we determine their local $\phi'(t)$[2] – reflecting how 'critical' t is to the timely completion of the workflow and is computed by removing those tasks from W which do not depend on t (to give W') and computing the new critical path $(CR(W'))$. Then ϕ' can be computed using Eq. 1 with $\phi' = p(W)$, $s(W) = s_t$ (the submission time of task t), $c(W) = d$ and $CR(W) = CR(W')$. Here ϕ' will be the proportional difference between the critical path and the time to the deadline d. Note that $\phi' \geq \phi$ with equality when t is part of the critical path for the whole workflow. The number of replicas to be run can now be determined through the use of RL which is parameterised by both the time of the day at which task t is submitted and the value of ϕ'.

– **Task completes:** We define task completion to be the time at which the first replica of a given task completes. First all other replicas are terminated as they are no longer of any use. We can then determine any new tasks for the workflow which are now free to be executed – those task for which all prior task have completed. For each of these tasks t we can compute ϕ' (as above) and hence the number of replicas which should be deployed at this time of day, until the workflow has completed.

4.2   Metrics

In order to evaluate the effectiveness of our approach and to provide metrics for use by our approach we define the following metrics:

– **Number of workflows with an overall $p(W) \leq P$:** Our users seek to have their workflows finish within the shortest time which is proportional to the length of the critical path of the workflow. We seek to identify the number of workflows which complete with a contingency less than P.

---

[1] Where the user desires $p(W) \leq \phi$.
[2] To simplify presentation we omit the parameters hereafter.

– **Energy consumption from replicas**: Here we report both the 'good' energy (that was used for the successful task) and the 'bad' energy for replica tasks. In the case of 'bad' energy this is the combination of both replicas which fail to be the first to complete and any replica runs which are terminated before completion. More formally for task i this is defined as:

$$bad_i = \sum_{r \in X_i} \sum_{j \in I_{i,r}} \begin{cases} \tau_{i,r,j} E_{i,r,j} & \text{if } G_{i,r,j} \neq 1 \\ 0 & \text{otherwise} \end{cases}, \qquad (3)$$

where X_i is the set of all replicas of the task i, $I_{i,r}$ is the set of all invocations of replica r, $\tau_{i,r,j}$ is the execution time of replica r of task i and invocation j, $E_{i,r,j}$ is the energy consumption rate (Watts) of the resource selected for replica r task i invocation j and $G_{i,r,j}$ is one for the good replica invocation, else zero. Likewise:

$$good_i = \tau_{i,r,j} E_{i,r,j}, \quad r \in X_i, j \in I_{i,r}, \text{ such that } G_{i,r,j} = 1.$$

Note that only one $G_{i,r,j} = 1$ and the total energy consumed for running the replicated task will be $good + bad$.

4.3 Computation of ϕ'

In order to compute ϕ' we must first determine the time remaining until the workflow deadline, which can be computed as $d - s_t$. Then, we estimate the time that the rest of the workflow will take to complete. The time to complete task t and all those tasks which are dependant on t, is computed as:

$$L_t = \begin{cases} r_t + \max_{j \in C_t} L_j & \text{if } |C_t| \neq 0 \\ r_t & \text{otherwise} \end{cases}, \qquad (4)$$

where C_t is the set of tasks immediately dependant on task t, and r_t is the duration of task t. We consider L_t to be the local critical path from task t to the end of the workflow. Note that for simplicity here we assume r_t includes any time for data ingress/egress and setup time. Note also that if t is the start task of the workflow then L_S is the duration of the critical path for the whole workflow.

Given that before workflow execution starts we do not know the values of r_t we can replace these with estimates (e_t) of the execution time. These estimates could be derived through a performance repository used to predict task execution times [15] or the use of performance prediction [28]. Equation 4 becomes:

$$L'_t = \begin{cases} e_t + \max_{j \in C_t} L'_j & \text{if } |C_t| \neq 0 \\ e_t & \text{otherwise} \end{cases}. \qquad (5)$$

The workflow will rarely follow this expected execution duration. Tasks may take more or less time than their estimate – either due to miss-estimates of their execution time or due to the fact that multiple submissions are required before certain tasks complete. This can lead to tasks, which weren't on the critical path, becoming part of the new critical path. Before submitting each task within the workflow we need to determine its potential impact on the overall workflow and how we can minimise this impact – say by running multiple copies of the task to increase its chance of completing within its own expectation time $((1 + \phi)e_t)$.

We have evaluated two approaches can be taken for determining the potential impact of a task on the overall workflow – both based around ϕ:

All Remaining Tasks Balanced: In this case we compute the spare time available for all task (including this task) which depend on this task and divide this proportionally between all these tasks. Thus we compute ϕ' as the proportion we can allocate to all remaining tasks that are dependant on task t:

$$\phi' = \frac{d - s_t}{L'_t} - 1.$$

With three cases for the value of ϕ':

- $\phi \leq \phi'$: Here, for this path of the workflow we are currently not part of the critical path, or we are on the critical path but ahead of schedule. There is currently no additional risk to completing the workflow by d. Running replicas of task t can be performed if deemed necessary by the time of day.
- $0 \leq \phi' < \phi$: We have fallen behind schedule for this path of the workflow, though there is still a probability that the workflow can be finished successfully. Remedial actions can be taken (such as running multiple copies of the current task) to increase the chance of completing the workflow by d.
- $\phi' < 0$: The workflow is significantly behind schedule and this path is likely to prevent the workflow from completing before d, remedial work is required. Note that if the remaining tasks in this path complete faster than their expected execution time it may still be possible to finish by d.

Balance on Current Task: Instead of balancing all remaining slack time across the remaining tasks in this path we assume all remaining tasks will still have their proportion ϕ while the current task will do anything possible to get itself back on track. We define ϕ'' as the proportion of extra time available to task t:

$$\phi'' = \frac{d - s_t - \max_{j \in C_t} L'_j}{e_t} - 1.$$

The three cases for ϕ' above also apply to ϕ'', though for $\phi'' < 0$ there is still slack time for the remaining tasks so the workflow may still finish on time.

4.4   Reinforcement Learning

We wish to determine for a given task t, submitted at time s_t, and with contingency proportion ϕ'^3 the number of replicas which should be submitted. We

---

[3] without loss of generality we use ϕ' to represent both ϕ', and ϕ''.

include here one as a valid number of replicas. In order to tailor our approach to
an individual HTC system we use Reinforcement Learning (RL) [33], to train an
agent to provide the number of replicas which is expected to give the greatest
reward (chance that the task will complete in the minimal time). RL is a form
of Machine Learning which can learn the 'best' action to perform given a particular system state. This can be achieved without training data – with training
coming from the interaction between an agent and a reward function which provides feedback on the actions taken by the agent. Thus RL can, not only, adapt
itself to any given environment but also, as it continually trains, adapt as the
environment changes. RL has been previously used to solve control problems
such as elevator scheduling, resource allocation within a data centre [3], reduction of energy consumption within volunteer computer systems [25] and bidding
stratergies for energy markets [18].

In order to use RL to optimise the number of replicas we use the approach
of an n-armed bandit [33]. Under this assumption each action – the number of
replicas to run – is independent of all other actions performed.

Each task $t \in \{1, 2, ...\}$ which we wish to replicate will observe the system in
a given state $s \in S$. Our state space here represents those characteristics of the
system over which decisions should be made – in this case the time of day at
which the task is to be submitted and the contingency proportion available for
the task. As these need to be discrete values we round the time of submission
to the hour of the day when submission took place, thus giving 24 states for
time. Likewise for contingency proportion we discretise this to $n + 2$ intervals.
Namely:

$$\{\phi' \leq 0, 0 < \phi' \leq \frac{P}{n}, ..., \frac{P(n-1)}{n} < \phi' \leq P, P < \phi'\}.$$

Thus our state space comprises of $24(n + 2)$ states. In order to maintain our
n-armed bandit model we assume that a task which has to relinquish a resource
becomes a new task within the system when it is re-allocated. The set of actions
$(a \in A)$ is the number of replicas which to be submitted to the system. We
determine the action a to perform as:

$$a = f(Q(s, A)), \tag{6}$$

where $Q(s, A)$ is the set of all reward values for the actions A available when the
system is in state s and $f()$ is a selection policy. The true reward values $Q(s, A)$
are unknown, but we can estimate $Q'(s, A)$ from the prior decisions which have
been made and the associated rewards. This becomes an estimator for $Q(s, A)$:

$$Q'_t(s, A) = \{q'_t(s, a)\} \quad \forall a \in A, \tag{7}$$

$$q'_t(s, a) = \overline{R_i(s, a')} \quad \forall i \leq t, a' = a, \tag{8}$$

where $R_t(s, a) \in [-k, k]$ is the reward for taking action a in state s for task t. A
value, for $R_t(s, a)$, of $-k$ indicates this was the worst possible choice of action
whilst $+k$ indicates the best choice of action. The value of k can be chosen
arbitrarily, however, it is normally a small number to prevent buffer overflows.
There are two possible outcomes when an action is applied to a task. These are:

- **Task t completed within the contingency proportion ϕ':** This is seen as a success for action a which should be rewarded thus increasing the chance of this action being selected again in the future. However, if left unchecked using just a reward for success here could lead to a system learning that the highest reward is obtained by allocating the maximum number of replicas to each task. We therefore need to diminish this reward proportional to the wasted work completed for those replica tasks which have been run.
- **Task t failed to complete within the contingency proportion ϕ':** This is seen as a failure for the action a chosen and requires a punishment to reduce the chances of this action being selected again in the future.

Therefore we can define the reward function as follows for task t:

$$R_t(s,a) = \begin{cases} +1 - \sigma_t & t \text{ completed within } \phi'e_t \\ -5 & t \text{ failed to complete within } \phi'e_t \end{cases},$$

where the first term in the reward function is used to indicate that the chosen action was good or not and the second term (if present) helps to steer the replication task towards the minimum value. We set the value for failed tasks to -5 to incur a large penalty for failure, the RL approach was not significantly affected by changing this value. We set the value of $\sigma_t \in [0,1]$ to be proportional to the wasted work performed by replicas. We compute σ_t using the energy consumption of the 'bad' replicas as a proxy for the wasted work:

$$\sigma_t = \min(1, \frac{bad_t}{ar_t \Xi}),$$

where Ξ is the average energy consumption rate for the selected resource when performing work – we assume 100% utilisation, a is the number of replicas and r_t is the actual execution time for the task. We include a in the denominator to scale with the bad_t value in the numerator. Otherwise the fraction would quickly become greater than one.

We can now define the selection policy $f()$ which is used to evaluate the action to perform given the prior history reward set $Q'(s, A)$. We define two approaches, a greedy (exploitative) and an explorative selection policy:

$$f(Q'(s,A)) = \begin{cases} max_a(Q'(s,A)) & \text{with probability } 1 - \epsilon \quad \text{(exploitative)} \\ random(A) & \text{with probability } \epsilon \quad\quad \text{(explorative)} \end{cases} \quad (9)$$

Here $max_a()$ selects the action a with the highest expected reward, whilst $random(A)$ will select an action uniformly at random from A.

By selecting the greedy policy we are exploiting prior knowledge to use the action with the greatest expected reward, whilst an exploitive policy allows us to search for potentially better actions. The dynamic and changing nature of our system necessitate both exploitative and explorative policies. Being too greedy can lead to poor choice of replica counts as the agent will keep using sub-optimal actions, whilst being too explorative can lead to the use of sub-optimal actions which are known to be bad. A careful selection of ϵ is therefore required.

Vary ϵ: In most cases, like ours here, the RL system will start off in an uninitiated state where each $Q'(s, a)$ has the same value (normally zero). If the value of ϵ is too small then the system can keep choosing the wrong action, when in exploitation mode, due to insufficient training. Likewise, if the underlying system changes the 'learned' actions may no longer be valid – in which case we should return to an explorative policy. We can therefore choose to vary ϵ during execution to allow better training. Two common approaches are:

- **Initially high ϵ:** Initially the value of ϵ is set high (ϵ_1), then to a lower value ϵ_2 after the first l rewards have been observed. This allows the RL to be initially more explorative and once the system has had a chance to 'learn' the best actions it will revert to a more exploitative policy.
- **Vary ϵ when results of choosing an action vary from those expected:** Here a sliding window captures the results from the last m selections of a given action. If the average reward of this value deviates too far then the value of ϵ can be increased for a time until it is deemed that the system has been re-trained. This allows the RL to become more explorative when the rewards move away from the expected range, to adapt to underlying changes.

4.5   Mechanisms of Workflow Task Enactment

In this work we compare three mechanisms for workflow task enactment:

- **Single task execution:** Each task within a workflow is submitted only once to the HTC system. This provides a baseline for workflow execution times.
- **Fixed replica execution:** The number of replicas is fixed. Although this can lead to reduced execution time of the workflow it has two main disadvantages. Firstly, tasks will be needlessly replicated at times of the day when this is not required. Secondly, the extra replicas could lead to contention for limited resources, hindering tasks which otherwise would have completed.
- **RL-based replica execution:** The number of replicas is determined dynamically at runtime. The likelihood of workflows finishing within the defined contingency is increased, whilst minimising overheads and contention.

5   Experimental Setup

We extend the HTC-Sim simulation [9,26] to model task replicas and incorporate RL functionality over workflows. We use our trace logs [9] from the use of the HTCondor [23] system and interactive users at Newcastle University during 2010. These logs represent 1,229,820 interactive user logins – the primary users of the cluster of 1,400 computers. During the year a total of 561,851 HTC tasks were submitted. As our original trace-log lacks examples of workflows, we randomly replace w% of the tasks within the trace-log with synthetic workflows. Here we choose w = 10% and use the same modified trace-log for all experiments to ensure consistency. We experiment with two workflow types: the simple workflow from Fig. 1 (each task taking 32 min), and Montage [17].

Table 1. Baseline case

ϕ	0.1	0.2	0.3	0.4	0.5	0.6	0.7	0.8	0.9	1.0	2.0
Successful	14,982	19,622	22,979	24,791	25,475	26,494	29,967	33,081	35,015	36,318	39,650
Energy	28.83	30.86	30.87	29.86	29.08	29.78	30.98	30.25	28.92	28.49	30.72

We have limited the maximum number of replicas, of each workflow task, to ten as experiments have shown that values greater than this give no advantage and often cause the cluster to become overloaded. Values of ϕ range between 0.1 and 1.0 in steps of 0.1 – we also evaluate the value of $\phi = 2$. For our state-space we have chosen $n = 10$ thus matching in with the intervals used for ϕ. We varied $\epsilon_1 \in [0.05, 1]$ and $\epsilon_2 \in [0.05, 0.4]$ and the sliding window $m \in [10, 10,000]$.

6 Results

We present here only the results for the "All remaining tasks balanced" case for brevity as we observed no statistical difference between the two approaches. We used $\epsilon_1 = 0.5, \epsilon_2 = 0.05$. Table 1 presents the base case, no task replication, indicating the number of workflows which completed within P contingency (Successful) and the energy consumed in these cases (Energy in MWh). The number of successful workload completions increases with P as expected. Energy values vary primarily due to the randomness within the simulation.

Figure 3a shows the increase in workflows which finish within a contingency proportional to P by using a fixed number of replicas, compared to our baseline case (Table 1). We see favourable results in the two- and three-replica case. This is particularly evident for contingency value $P = 0.1$, where over 20% more workflows are capable of meeting the target. Four or more replicas lead to contention for resources and corresponding performance degradation. However, if the number of replicas can vary, then it may be beneficial to deploy more than four replicas at specific times of the day or for particular contingency levels.

Figures 3b and 3c indicate the extra energy required when we run a fixed number and RL chosen replica count for each task within a workflow. As can be seen from this figure the energy consumption increases linearly with the increase in replicas (for the RL case this is the maximum number of replicas it can select). As such this will quickly lead to excessive extra computation and wasted work. However, the RL case effectively halves the energy consumption.

Figure 4 shows how much we can reduce the energy consumption of running replicas by allowing RL to select the number of replicas to enact. For fair comparison, RL is bounded by the same maximum number of replicas as the fixed case. The percentage increase in energy is presented in Fig. 3c. We see our approach achieves promising energy savings for all replication scenarios (replica count $\geq$ 2), successfully identifying system conditions requiring the use of replication, while minimising unnecessary replication during periods of low utilisation which would lead to wasted energy. Although we can significantly reduce the energy consumption of enacting replicas, we must also consider the performance impact

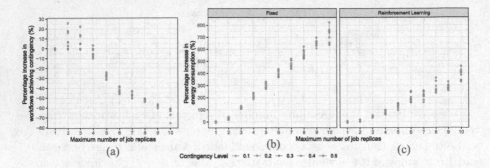

Fig. 3. For fixed replicas: a) Percentage of workflows achieving users desired contingency. b) Energy requirements for achieving users desired contingency. c) Percentage increase in energy consumption by using RL.

of this additional replication. Figure 5 illustrates the proportion of workflows able to complete within time contingency P. We observe favourable improvement in workflow completion, particularly for small contingency values. Importantly, we see improvements across a much wider range of replica counts. This shows our approach is far less susceptible to sub-optimal replica count selection.

We now validate our approach for the well-established Montage workflow [17]. Figure 6a shows our approach can improve the number of successful workflows. This seems to be irrespective of the choice of replica count, however, this is most likely a consequence of the newly successful workflows only using a small number of replicas. The biggest impact comes from the choice of contingency – with lower contingency giving better results. Likely a consequence of RL not over-provisioning for cases where it can't complete on time, thus keeping the system idle for other workflows. Figure 6b confirms that the energy impact is reduced significantly compared to a fixed replication scheme. We are able to conserve 34% energy consumption with only a 4% decrease in workflows achieving contingency.

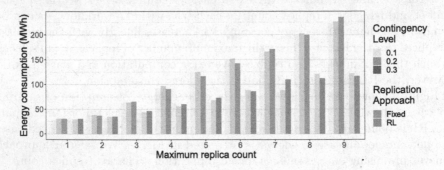

Fig. 4. Comparison between energy increase for fixed replicas and RL replicas

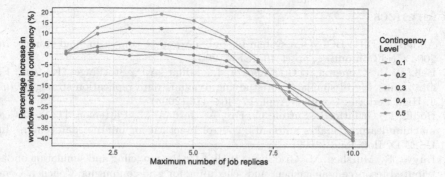

Fig. 5. Increase in number of workflows which can be performed when using RL

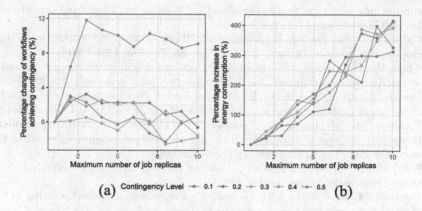

Fig. 6. Using RL for Montage. **a)** Percentage increase in workflows achieving users' desired contingency. **b)** Percentage change in energy consumption.

7   Conclusions

We have explored, through trace-driven simulation, how the performance and energy consumption of workflow enactments in an HTC environment can be improved through task replication. We demonstrate that fixed replication schemes, if tuned correctly, can deliver significant performance improvements, but impose a considerable overhead in energy consumption. In contrast, our proposed Reinforcement Learning approach curtails the energy consumption while retaining the performance benefits. We show our approach to be less susceptible to poor performance due to sub-optimal replica count selection.

Future work will explore the broader applicability of RL approaches in workflow enactment; e.g. an RL-based overlay to the HEFT scheduling heuristic.

References

1. Anderson, D.P.: BOINC: a system for public-resource computing and storage. In: 2004 Grid Computing, pp. 4–10. IEEE (2004)
2. Bell, W.H., Cameron, D.G., Capozza, L., Millar, A.P., Stockinger, K., Zini, F.: Optorsim - a grid simulator for studying dynamic data replication strategies. Int. J. High Perform. Comput. Appl. **17**, 403–416 (2003)
3. Bodík, P., Griffith, R., Sutton, C., Fox, A., Jordan, M., Patterson, D.: Statistical machine learning makes automatic control practical for internet datacenters. In: USENIX HotCloud (2009)
4. Buyya, R., Murshed, M.: GridSim: a toolkit for the modeling and simulation of distributed resource management and scheduling for grid computing. Concurr. Comput. Pract. Exp. **14**(13), 1175–1220 (2002)
5. Buyya, R., Ranjan, R., Calheiros, R.N.: Modeling and simulation of scalable cloud computing environments and the CloudSim toolkit: challenges and opportunities. In: HPCS 2009, pp. 1–11. IEEE (2009)
6. Calheiros, R.N., Buyya, R.: Meeting deadlines of scientific workflows in public clouds with tasks replication. IEEE TPDS **25**(7), 1787–1796 (2014)
7. Chen, W., Deelman, E.: WorkflowSim: a toolkit for simulating scientific workflows in distributed environments. In: IEEE e-Science 2012, pp. 1–8. IEEE (2012)
8. Durillo, J.J., Nae, V., Prodan, R.: Multi-objective workflow scheduling: an analysis of the energy efficiency and makespan tradeoff. In: IEEE/ACM CCGrid (2013)
9. Forshaw, M., McGough, A., Thomas, N.: HTC-Sim: a trace-driven simulation framework for energy consumption in high-throughput computing systems. Concurr. Comput. Pract. Exp. **28**(12), 3260–3290 (2016)
10. Forshaw, M.: Operating policies for energy efficient large scale computing. Ph.D. thesis, Newcastle University, UK (2015)
11. Forshaw, M., McGough, A.S., Thomas, N.: Energy-efficient checkpointing in high-throughput cycle-stealing distributed systems. Electron. Notes Theor. Comput. Sci. **310**, 65–90 (2015)
12. Forshaw, M., Thomas, N., McGough, A.S.: The case for energy-aware simulation and modelling of internet of things (IoT). In: Proceedings of the 2nd International Workshop on Energy-Aware Simulation, pp. 5:1–5:4. ENERGY-SIM, ACM (2016)
13. Forshaw, M., Thomas, N., McGough, S.: Trace-driven simulation for energy consumption in high throughput computing systems. In: IEEE/ACM DS-RT (2014)
14. Georgakopoulos, D., Hornick, M., Sheth, A.: An overview of workflow management: from process modeling to workflow automation infrastructure. Distrib. Parallel Databases **3**(2), 119–153 (1995)
15. Hiden, H., Woodman, S., Watson, P.: A framework for dynamically generating predictive models of workflow execution. In: Proceedings of the 8th Workshop on Workflows in Support of Large-Scale Science. WORKS 2013 (2013)
16. J. E. Kelley, J., Walker, M.R.: Critical-path planning and scheduling. In: International Workshop on Managing Requirements Knowledge, p. 160 (1959)
17. Jacob, J.C., et al.: Montage: a grid portal and software toolkit for science-grade astronomical image mosaicking. Int. J. Comput. Sci. Eng. **4**(2), 73–87 (2009)
18. Kell, A.J.M., Forshaw, M., Stephen McGough, A.: Exploring market power using deep reinforcement learning for intelligent bidding strategies. In: 2020 IEEE International Conference on Big Data (Big Data), pp. 4402–4411 (2020)
19. Kliazovich, D., Bouvry, P., Audzevich, Y., Khan, S.U.: GreenCloud: a packet-level simulator of energy-aware cloud computing data centers. In: GLOBECOM (2010). https://doi.org/10.1109/GLOCOM.2010.5683561

20. Legrand, A., Marchal, L.: Scheduling distributed applications: the simgrid simulation framework. In: Proceedings of the Third IEEE International Symposium on Cluster Computing and the Grid, pp. 138–145 (2003)
21. Li, Z., et al.: A security and cost aware scheduling algorithm for heterogeneous tasks of scientific workflow in clouds. Future Gener. Comput. Syst. **65**, 140–152 (2016)
22. Lim, S.H., Sharma, B., Nam, G., Kim, E.K., Das, C.: MDCSim: a multi-tier data center simulation, platform. In: 2009 IEEE International Conference on Cluster Computing and Workshops, CLUSTER 2009, pp. 1–9 (2009)
23. Litzkow, M., Livney, M., Mutka, M.W.: Condor-a hunter of idle workstations. In: ICDCS (1988)
24. McGough, A.S., Forshaw, M., Gerrard, C., Wheater, S.: Reducing the number of miscreant tasks executions in a multi-use cluster. In: 2012 Second International Conference on Cloud and Green Computing (CGC), pp. 296–303 (2012)
25. McGough, A.S., Forshaw, M.: Reduction of wasted energy in a volunteer computing system through reinforcement learning. Sustain. Comput. Inform. Syst. **4**(4), 262–275 (2014)
26. McGough, A.S., Forshaw, M.: Energy-aware simulation of workflow execution in high throughput computing systems. In: IEEE/ACM DS-RT (2015)
27. McGough, A.S., Forshaw, M.: Evaluation of energy consumption of replicated tasks in a volunteer computing environment. In: Companion of the 2018 ACM/SPEC International Conference on Performance Engineering, ICPE 2018, pp. 85–90 (2018)
28. McGough, A.S., Afzal, A., Darlington, J., Furmento, N., Mayer, A., Young, L.: Making the grid predictable through reservations and performance modelling. Comput. J. **48**(3), 358–368 (2005)
29. McGough, A.S., Forshaw, M., Gerrard, C., Robinson, P., Wheater, S.: Analysis of power-saving techniques over a large multi-use cluster with variable workload. CCPE **25**(18), 2501–2522 (2013). https://doi.org/10.1002/cpe.3082
30. Méndez, V., García, F.: SiCoGrid: a complete grid simulator for scheduling and algorithmical research, with emergent artificial intelligence data algorithms (2005)
31. Niu, S., et al.: Employing checkpoint to improve job scheduling in large-scale systems. In: Cirne, W., Desai, N., Frachtenberg, E., Schwiegelshohn, U. (eds.) JSSPP 2012. LNCS, vol. 7698, pp. 36–55. Springer, Heidelberg (2013). https://doi.org/10.1007/978-3-642-35867-8_3
32. Roy, A., Livny, M.: Condor and preemptive resume scheduling. In: Nabrzyski, J., Schopf, J.M., Węglarz, J. (eds.) Grid Resource Management. International Series in Operations Research & Management Science, vol. 64, pp. 135–144. Springer, Boston (2004). https://doi.org/10.1007/978-1-4615-0509-9_9
33. Sutton, R., Barto, A.: Reinforcement Learning: An Introduction. A Bradford book, Bradford Book (1998)
34. Tannenbaum, T., Wright, D., Miller, K., Livny, M.: Condor: a distributed job scheduler. In: Beowulf cluster computing with Linux, pp. 307–350. MIT press (2001)

35. Topcuoglu, H., Hariri, S., Wu, M.Y.: Performance-effective and low-complexity task scheduling for heterogeneous computing. IEEE Trans. Parallel Distrib. Syst. **13**(3), 260–274 (2002)
36. Yu, J., Buyya, R., Ramamohanarao, K.: Workflow scheduling algorithms for grid computing. In: Xhafa, F., Abraham, A. (eds.) Metaheuristics for Scheduling in Distributed Computing Environments. Studies in Computational Intelligence, vol. 146, pp. 173–214. Springer, Heidelberg (2008). https://doi.org/10.1007/978-3-540-69277-5_7
37. Zhu, Q., Zhu, J., Agrawal, G.: Power-aware consolidation of scientific workflows in virtualized environments. In: Proceedings of the ACM/IEEE International Conference for High Performance Computing, Networking, Storage and Analysis (2010)

Performability Requirements in Making a Rescaling Decision for Streaming Applications

Paul Omoregbee[ID] and Matthew Forshaw[✉][ID]

Newcastle University, Newcastle Upon Tyne, UK
{paul.omoregbee2,matthew.forshaw}@newcastle.ac.uk

Abstract. Maximising the benefits of auto-scaling is difficult due to challenges associated with precisely estimating resource usage in the face of significant variability in client workload patterns and trends. Most auto-scaling systems do not consider the impact of scaling time and often assume that application state sizes and offered load remain unchanged during the scaling interval. Long scaling durations, especially in rapidly varying workload environments, pose a challenge to auto-scaling, exacerbating the impact of suboptimal parallelism choices. In this paper, we empirically evaluate application state size and end-to-end checkpoint duration and identify their correlation with the duration of scaling procedures. We then develop predictive models to provide future autoscalers with further intelligence to inform scaling decisions.

Keywords: Auto-scaling · Rescaling duration · State size

1 Introduction

Stream processing technology powers numerous applications, such as continuous analytics, monitoring, fraud detection, stock trading, and mobile and network information management [1]. Dealing with applications characterised by high demand fluctuation in stream processing is difficult, as the distribution of data streams is constantly changing and unpredictable [2].

Workload variance and skewness are common features in distributed stream processing deployments. When a large amount of data is injected into a distributed system for processing, a change in the source data stream can affect the system performance resulting in Service Level Objective (SLO) violation [3]. Cloud hosting enables operators to scale horizontally or vertically based on the benefit of on-demand elasticity and economy of scale. However, without sufficient knowledge of the workload characteristics, making a scaling decision can become a complex task [4]. This could lead to over-provisioning or under-provisioning. Overprovisioning system resources can lead to increased costs and the underutilisation of computational resources. Meanwhile, under-provisioning can lead to suboptimal performance of the system, thereby affecting the Quality of Service (QoS). Auto-scaling is the ability to automatically adjust a system capacity or compute resource to maintain a steady or predictable performance level at the lowest possible cost.

K. Gilly and N. Thomas (Eds.): EPEW 2022, LNCS 13659, pp. 133–147, 2023.
https://doi.org/10.1007/978-3-031-25049-1_9

This paper considers the impact of long rescaling duration during an auto-scaling interval which could lead to multiple rescaling of an application when the state size growth of the application is larger and unpredictable. State size is the measure of the entire content of the memory where the application state resides at a given point in time.

Figure 1 shows the estimated time to auto-scale an application over different state sizes. The plot shows that it takes approximately 40 s to scale a running application when the application state size is ~ 1 GB. In this experiment, we collect the application state information twice, every 2.5 min. Each shape represents the duration of the sliding window; consider each shape as the scaling duration value collected every 2:30 min for the different state sizes. Therefore, we have two identical shapes for every deployment (each experimental deployment is set at 5 min). You will observe that the points begin to disperse as the state size grows. This shows the rescaling time variance between each deployment.

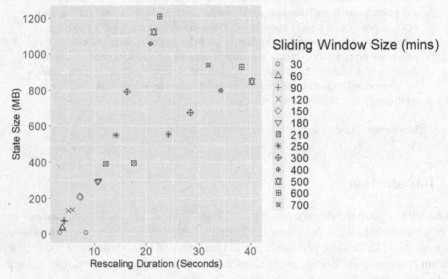

Fig. 1. Rescaling duration increasing over larger state sizes

Figure 1 shows that, as the application's state grows, so does the rescaling duration. We argue that more data could have accumulated during a rescaling process, which could mean constantly rescaling and falling short of resources, leading to multiple rescaling of the application. This repetitive task could harm system performance. We believe that a proactive approach will be to use machine learning algorithms to develop a workload prediction module that can forecast workload characteristics [5] and rescaling duration. This predictive approach should be embedded in a scaling policy to broaden the decision-making scope and ensure adequate resource allocation is made even when workload skewness exists.

We make the following contributions in this paper. First, we show that rescaling duration is critical in auto-scaling a streaming application. Second, we show that state size is an important metric to consider when making a rescaling decision. Third, we show that the end-to-end checkpoint duration strongly correlates with the state size. Fourth, we develop a machine learning predictive model to forecast a streaming application's rescaling duration based on the state size and end-to-end duration. This model will provide a scaling controller with knowledge of the estimated rescaling duration. Finally, we apply our model to predict the rescaling duration of an application based on forecasted state size values.

This paper is organised as follows. Section 2 provides the motivation, background, and related work; in Sect. 3, we provide details of the system design approach; Sect. 4 provides details of our trained predictive models used to predict rescaling duration based on state size; Sect. 5 provides a summary of our experimental results and Sect. 6 provides concluding remarks and some questions for future work.

2 Background and Related Work

Long-running applications inevitably face various external and internal shocks capable of creating instability, for example, users' reactions to a national disaster on Twitter or a champion's league football goals. Unpredictable load variation can lead to over-provisioning or under-provisioning.

Prior research has introduced automatic rescaling controllers [6–8]. A key service distinction is the approach taken by various systems to react to changes in real-time and balance competing objectives like performance (throughput or latency) and resource utilisation during a load spike or resource failure [9, 10].

Some research has argued that memory and CPU utilisation metrics and other coarse-grained metrics are inadequate for arriving at a good scaling decision, especially in a mul-titenancy cloud environment where shared resources and performance interference are prevalent [7, 11]. Research is increasingly focusing on operator performance, dataflow topology, and the sustainability of throughput levels at different cluster sizes for different streaming frameworks [6, 12, 13]. We take a similar approach for our horizontal scaling experiments, investigating the operator throughput capacity. However, this paper focuses on the performability requirement in rescaling when an application state size becomes bigger.

A suitable scaling controller should provide Stability, Accuracy, Short Settling Time, and Overshooting (SASO) properties [7]. Whereas speculative scaling techniques that violate these could lead to unnecessary costs due to sub-optimal utilisation of resources due to over-provisioning or under-provisioning, performance degradation due to frequent scaling action due to oscillation and low convergence resulting in an SLO violation or load shedding [7]. Furthermore, an effective scaling controller must be mindful of the resource availability and the scaling duration. These factors are an important element that must be contained in the scaling policy.

Mindful of the challenges that come with decision-making on how to scale efficiently and when to scale, we note the work carried out in the DS2 project, which provides an automatic scaling system that strives to strike a balance between resource over-provisioning and on-demand scaling.

DS2 evaluates each operator's true and observed processing capabilities regardless of the backpressure and other effects. True processing rate signifies the maximum potential number of records an operator instance can process per unit of useful time (duration minus waiting time). Intuitively, this calculates the capacity of operator hardware. The observed processing rate is the number of records an operator instance processes per unit of observed time (duration plus waiting time) [7].

Based on real-time performance traces, DS2 automatically determines, on demand, the optimal level of parallelism for each operator in the dataflow. The number of resources allocated to each operator is maintained as a dynamic provisioning plan. In this experiment, we adopted the default DS2 policy configuration values when applied to Apache Flink: 10-s decision interval (frequency of metrics collection and policy application), 30-s warm-up period (the number of consecutive policy intervals that were ignored following a scaling action), one interval activation period (when DS2 will make a scaling decision), and a 1.0 target ratio (maximum permitted difference between the target rate and observed source rate attained by the policy).

However, like DS2, most auto-scaling systems do not consider the impact of scaling duration, especially in a volatile workload environment. Instead, more attention is focused on mitigating the impact of over-provisioning of resources for temporary load spikes and under-provisioning during peak loads.

2.1 Influencing Metrics for Scaling a Streaming Application

Scaling controllers usually rely on metrics to determine when and how to scale. Several studies have focused on different metrics in building an optimal scaling policy. Most systems rely on simplistic performance models, like setting predefined thresholds and conditions. These conditions include CPU and memory utilisation, backpressure, observed processing rate, etc. However, CPU and memory utilisation can be insufficient metrics for streaming applications, especially in cloud environments where multi-tenancy and performance interference is prevalent [9].

Kalavri et al. propose a better approach (DS2) to automatically determine the optimal parallelism level for each operator in the dataflow as the computation progresses, using real-time performance traces. It maintains a dynamic provisioning plan in which the allocation of resources to each operator changes.

This paper explores other metrics like state size and end-to-end checkpoint duration. These metrics have been shown to significantly influence the scaling duration of streaming applications compared to offered load (data arrival rate).

3   System Design

Figure 2 shows our experimental pipeline setup and the relationship among the various unit of our setup. These units are self-contained, which makes it easy to scale the experiment. This setup architecture has four major areas: the data source (NEXMark workload generation), Streaming framework (Flink and DS2 scaling controller), Visualisation, and Datastore. Our experimental aim is to simulate a fluctuating workload streaming environment and measure the time required to complete a rescaling procedure using the DS2 rescaling policy. This experiment leverages Flink's consistent checkpointing of application state on a single Flink instance [13].

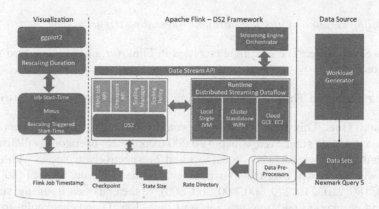

Fig. 2. Experimental system architecture

Hardware and Software Configuration. This experiment was run on Ubuntu `4.15.0-74-generic` with a single instance of Flink version 1.4.1. The hardware configuration includes `Intel® Core™ i5-8500 CPU @3.00GHz` (six cores), 16 GB of memory, and 1 TB of storage.

We chose the file system state backend (`FsStateBackend`). Further work would merit exploring the differing impacts of alternative state backends, such as `MemoryStateBackend` or `RocksDBStateBackend`.

NEXMark Query 5 was adopted as the workload for this experiment because of its ability to retain the system state based on the sliding window capabilities that meet our experimental objective to simulate the rescaling duration of a streaming application that accumulates a huge state. A sliding window groups tuples within a window that moves across the data stream in accordance with a predetermined interval. NEXMark is an online auction system that allows queries over three main business objects (Person, Auction and Bid) [2, 14].

Query 5 selects the items that have received the most bids over a period. These items are referred to as `HOT_ITEMS`. This query uses a sliding window group by operation. In sliding window aggregation, the application summarises a collection of recent streaming data using historical and current events [15].

We set the data source rate at 20,000, the checkpoint interval to 2:30 min and the sliding window to 60 min. We update the time window configuration parameter using different time intervals to create different state sizes each time we run the workload. Subsequently, we consume the `/jobs/:jobid/` REST API and collect values like state size and end-to-end checkpoint duration (this is the duration of a complete checkpoint measured by the time interval between the triggered timestamp and the most recent acknowledgement.) and job start time. Below is a summary of the two major experimental runs used for this experiment.

The purpose of running two experiments was to get the one that simulated our experimental objective better. We aim to show a use case where state size growth will lead to an increased rescaling duration interval.

Rescaling duration is the time between when a scaling operation is triggered, and the new configuration is deployed. We derive the rescaling duration as follows:

Rescaling duration = new job start time – rescaling triggered time.

New Job Start Time. The start time recorded in Flink for a newly deployed running application.

Rescaling Triggered Time. The recorded time when a rescaling procedure is triggered.

Experiment One. This experiment was repeated 13 times. Checkpoint interval equal to five minutes, data source rate equal to 20,000 records/sec, operator parallelism (Bid-Source and Sliding window-sink) equal to 1 and 2 respectively, windows width time in minutes were alternated over the 13 experiments in the following interval (30, 60, 90, 120, 150, 180, 210, 240, 270, 300, 400, 500, 600). At the end of each run, the state size, end-to-end checkpoint duration, new job start-time, Job ID and checkpoint triggered-start-time values are collected through the Flink APIs. We control the state size at the end of each deployment for each deployment by adjusting the window width time value of the workload.

Experiment Two. We try to explore shorter checkpoint intervals to enable us to measure the scaling duration between different checkpoints. Therefore, the checkpoint interval for this experiment is set to 2:30 min to allow us to collect the state size twice after each checkpoint and redeployment. Like the previous experiment, we adjusted the state size after every deployment through the window width time parameter of the workload. After each deployment, we update the sliding window-sink operator parallelism to simulate a change in the data flow topology during scaling.

This paper provides decision support and predictions that can cater for the early deployment of newly emerging workloads with no historical operational data. Therefore, the volume of data used for this experiment supports this use case. More experiments could be run with wider checkpointing intervals. However, this would require more system resources.

We also observed the backpressure status, which becomes high when the state size grows. This experiment measures the rescaling duration when the state size grows huge. Therefore, we do not investigate the backpressure effect because it affects the offered load, which is not a priority for this experiment.

We also kept the source data rate static (20,000 records/sec) as increasing it did not significantly affect the state size as we wanted. The choice of checkpoint interval (2:30 & 5 min) enabled us to collect a state size value distinctive from another. Shorter intervals would have very close results, while longer intervals would require a bigger hardware resource.

3.1 Impact of Long Checkpointing Duration on Streaming Applications

Flink checkpointing mechanism guarantees exactly-once state consistency and is the most common fault tolerance and rescaling approach used by most state-of-the-art streaming systems for stateful applications[16]. We measure the end-to-end duration of each checkpoint, which is the duration of a complete process.

Checkpointing intervals are usually carefully selected as high intervals could lead to longer recovery duration, while low intervals can lead to high processing overhead due to state size [13]. Figure 3 shows that a larger state leads to longer checkpoint durations, which becomes a potential danger for applications that require frequent checkpoints.

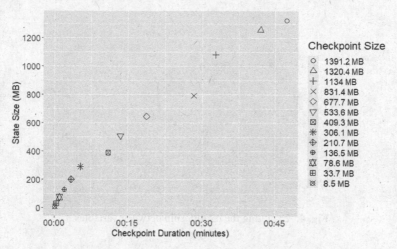

Fig. 3. Flink checkpoint duration increasing over larger state sizes

4 Predictive Modelling

This section describes our approach to developing our predictive models and the methods used in testing each model's performance quality. We also show the prediction results for each model when they are applied to known and unknown data sets.

4.1 Linear Regression Analysis for Predictive Modelling

To determine whether we can create a predictive model, we start by determining whether our predictor and response variables are related.

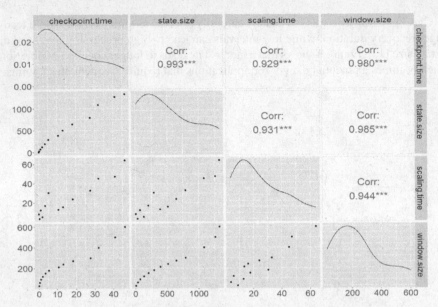

Fig. 4. Exploring the relationship between variables.

Figure 4 contains four variables from our dataset: checkpoint time (the end-to-end checkpoint duration for a complete checkpoint measured in seconds), state size (the state size of all acknowledged subtasks measured in megabytes), scaling time (this is the rescaling duration measured in seconds), and window size (the different sliding window configuration parameters used for each deployment we ran to produce different state sizes measured in minutes). The relationship between state size, end-to-end checkpoint duration, window width size and rescaling duration appears to have a strong relationship with correlation coefficients close to one. We can see a consistent increase in the rescaling time as the state size, end-to-end duration, and window width size increase. We formulate the equation for our linear regression as follows.

$$Y \approx \beta 0 + \beta 1 X + \varepsilon$$

- The Y and X represent the dependent and independent variables.
- $\beta 0$ denotes the model intercept or the point at which it crosses the y-axis.
- $\beta 1$ denotes the model slope and the direction and steepness of the line (positive or negative).
- ε is the error team that includes variability that the model cannot account for (what X is unable to reveal about Y).

To ascertain the strength of these relationships and identify which independent variables have the closest relationship with our dependent variable, we developed four linear regression models.

1. Model 1. *Rescaling Duration* ≈ *β0 + β1 (State Size) + ε*

2. Model 2. *Rescaling Duration* ≈ *β0 + β1 (State Size * End-to-End Duration) + ε*
3. Model 3. *Rescaling Duration* ≈ *β0 + β1 (State Size * Window Width Size) + ε*
4. Model 4. *Rescaling Duration* ≈ *β0 + β1 (State Size * End-to-End Duration * Window Width Size) + ε*

Since we have numeric values and the relationship tends to have a linear pattern, we choose to use the linear regression model for this experiment. We create a linear model engine to fit our linear model line to our data.

4.2   Resampling

We split the dataset into a training set and a testing set. We took the random sample approach to randomly select 80% of our dataset as training data set and the remaining 20% as test data. We note the work by Joseph et al. that proposes some evaluation criteria for choosing an optimal splitting ratio [17]. However, there is currently no consensus regarding the optimal data splitting ratio based on theoretical and numerical investigations [18, 19].

Table 1 below shows the comparison of all our models' statistical properties. Working through the output of our models' statistical properties, we focus on a few evaluation metrics that tell us how well these models fit our data.

Table 1. Models statistical properties

	Model 1			Model 2			Model 3			Model 4		
	95%	CI	p-value	95%	CI	p-value	95%	CI	p-value	95%	CI	p-value
State size	0.02	0.03	<0.001	0.00	0.02	0.077	0.00	0.03	0.10	−0.07	0.03	0.4
End-to-end duration				1.1	2.2	<0.001				−2.2	1.9	>0.9
State size * end-to-end duration				0.00	0.00	0.032				0.00	0.01	0.2
State size * end-to-end duration							0.06	0.18	<0.001	−0.03	0.20	0.2
State size * window size							0.00	0.00	0.003	0.0	0.0	>0.9
End-to-end duration * window size										0.00	0.01	0.3
State size * end-to-end duration * window size										0.00	0.00	0.024

Confidence Interval (CI) / Confidence level (95%). Sample size impacts the CI and confidence level estimate. Larger sample sizes normally lead to higher confidence levels.

A confidence interval with zero value may indicate no significance in the parameter's relationship. This is frequently how confidence intervals are interpreted, but this could be incorrect, as shown by other relevant statistical properties. Rather, it indicates uncertainty. Having a confidence interval of zero indicates that an independent variable could have a positive (> 0) or negative (< 0) effect on the rescaling duration. However, considering the volume of our sample size, we focus on other statistical metrics like the R-squared, Adjusted R-squared and p-value.

P-value. Statistical significance is defined as a p-value less than 0.05 (< 0.05). It indicates strong evidence against the null hypothesis, representing the likelihood that our data would have occurred under the null hypothesis. As a result, the null hypothesis is rejected, and the alternative hypothesis is accepted.

Table 2. Comparison of Model Performance Indices.

Name	R_2	R_2 (adj.)	RMSE	Sigma	AIC weights	BIC weights
Model 1	0.635	0.615	7.076	7.458	<0.001	<0.001
Model 2	0.937	0.925	2.937	3.283	0.158	0.579
Model 3	0.843	0.813	4.644	5.193	<0.001	<0.001
Model 4	0.964	0.944	2.212	2.855	0.842	0.421

Table 2 evaluates the performance quality of each model. Model 4 has the better performance indices results, followed by Model 2. Figure 5 gives a pictorial representation of the model performance indices comparison.

R-squared (R_2) is the correlation between the known outcome values and the model-predicted values. Values closer to 1 indicate models with a good fit; Adjusted R-squared R_2 (adj) is usually lower than or equal to R_2. Similarly, A model with accurate predictive capability has a value of 1, while a model with weak or no predictive value has a value of 0 or less; Calculating the Root Mean Square Error (RMSE) is one way to determine how well a regression model fits a dataset, by measuring the average distance between the predicted values of the model and the actual values in the dataset. The lower the root mean square error a model has, the better it fits a dataset; Sigma (standard deviation) quantifies the degree to which a process deviates from ideal performance. Furthermore, for model selection, Akaike Information Criterion (AIC) and Bayesian Information Criterion (BIC) are frequently utilized. AIC picks the best-fit model that explains the most variance with the fewest number of independent variables, while BIC assesses how likely a model is to be accurate. Lower AIC and BIC scores are preferable.

Further comprehensive model checks were carried out to check each model's assumptions like collinearity, normality of residuals, linearity, homogeneity of variance and influential observation. The result indicates a multicollinearity issue in Models 2, 3 and 4, with each having a variance inflation factor (VIF) above 10. Concerning the normality of residual, models 2 and 4 are normally distributed, result on linearity showed that model 2 data points are fitted closer to the residual line compared to other models, which suggests that it has a better predictive performance. The result of influential observation shows that model 4 and 2 has the highest leverage and residual, which indicates a strong influence on the observation by these two models.

Model 4 and Model 2 both stand out to be the most efficient for our experiment. However, we shall investigate further to get the best amongst the two.

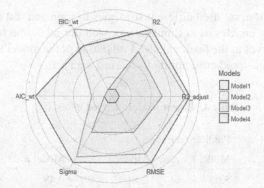

Fig. 5. Comparison of model indices.

Consequently, Cross-Validation (CV) is applied to our training data, which splits the training data further into the assessment and analysis set. An assessment set in the tidymodel framework is a set of data to measure the CV's performance, while analysis data sets are used to train and fit the models [20]. The overall objective is to measure our model's performance further using some performance statistics. Due to the volume of our data, we created a 4-folds CV. This randomly allocates the 20 cells in the training data set into four groups ("folds") of equal sizes.

For the initial iteration of resampling, the first fold of approximately five cells is held out for performance measurement, while the remaining 75% of the data (approximately 15 cells) is used to fit the model. After training the models on the analysis set, the four models are applied to the assessment set to generate predictions. Next, we compute the performance statistics for each model based on the predictions' results. Table 3 shows the resampling results and performance statistics created from the 4-fold cross-validation. The performance statistics used are the RMSE and the R_2.

Table 3. Performance statistics metrics for each model

Models	Model 1		Model 2		Model 3		Model 4	
Metrics	RMSE	R_2	RMSE	R_2	RMSE	R_2	RMSE	R_2
Mean	7.690	0.707	3.430	0.977	5.720	0.791	7.54	0.818
Standard error	0.309	0.077	0.635	0.012	0.794	0.022	1.22	0.041

Model 2 appears to be the most efficient model considering it has the lowest RMSE mean value and the highest R_2 mean value. However, if subjected to a larger data set, Models 4 and 2 could compete for superior quality performance.

5   Summary of Experimental Results

In the previous section, we used different strategies to train and test the efficacy of our models. Resampling enables us to simulate how well our model performs on new data, while the test set serves as the final, unbiased validation of the model's performance. We now apply the model's predictive ability to our test data set.

Table 4. Linear models applied to test data

Rescaling duration	Predicted duration			
	Model 1	Model 2	Model 3	Model 4
3.75	5.65	2.81	0.49	**3.29**
4.91	7.94	**4.74**	8.03	6.61
5.71	7.99	**6.13**	8.04	6.88
8.30	**5.02**	2.57	−3.36	1.36
14.17	18.10	15.50	20.10	**15.00**

Table 4 shows the result of predicted results by each model placed side by side with the actual values of the test data set. Values in the "Rescaling Duration" column are the known test data set aside when we carried out the 80/20 split operation. The predicted rescaling duration values are the predicted values generated by each model. The predicted results (bolded) are the closest predicted values produced by each model. Following these prediction results, Models 2 and 4 have a better predictive performance capability than other models.

State Size Forecasting. We apply our model to a real-world use case, aiming to show the general applicability of our model to common use cases. We leverage experimental data measurements from the Rule-Based Event Aggregator (RBEA) [21].

We extracted measurements of five distinct RBEA deployments, spanning from 100 to 500 GB of global state. The end-to-end duration was set to 9 ms based on empirical knowledge. We applied our Model 2 to estimate the rescaling duration for each state size measurement.

The results (83 s, 149 s, 215 s, 281 s and 347 s) show a linear growth in the rescaling duration that correlates with state size. Based on the result, it will take between 83–347 s to auto-scale the RBEA global state size.

We also measured Model 2's mean prediction values of rescaling duration over the forecasted state size values with the upper and lower limits prediction limits. We observe that as the state size grows bigger, so do the upper and lower limit prediction values. An estimate will always have a level of uncertainty associated with it, which is dependent on the underlying variability of the data as well as the sample size. The more variable the data, the greater our estimate's uncertainty. Similarly, as the sample size increases, we gain more information and thus reduce our uncertainty [22]. A limitation of the current modelling approach is the prediction of negative values. These negative predictions may be caused by the speculative approach of manually allocating corresponding end-to-end duration to match each forecasted state size. This can be explored in future work to prevent the model from predicting negative values.

6 Conclusion

Our experimental findings highlight the importance of state size during rescaling, demonstrating that with the accumulation of more state, the time to auto-scale also increases. More state could have accumulated during a rescaling time window, and this could mean constantly rescaling and falling short of resources which will lead to multiple rescaling of the application. This repetitive task could harm system performance.

Four predictive regression models were developed and trained with the existing data set, leveraging resampling cross-validation. Model 2 was chosen as the most efficient based on the performance statistics and predictive performance, followed by Model 4.

Next, we use Model 2 to predict the rescaling duration of a running application over some forecasted state size values. Results show that it will take approx. 6 min to auto-scale when the application state size reaches 500 GB. In a constantly changing and unpredictable distributed data streaming environment, a lot could happen in 6 min. We, therefore, recommend that forecasting workload characteristics and state size growth rate is very relevant when developing a scaling policy to enable a streaming application to handle unanticipated resource demand.

Some interesting future work includes developing an adaptive approach leveraging our models in this paper within DS2's scaling policy to govern rescaling decisions.

It will be interesting to evaluate the performance impact of a long-running checkpoint process on the streaming application with respect to overlapping checkpointing operations without delaying subsequent scheduled checkpoints [23].

A limitation of the current modelling approach is that it predicts negative values in the mean prediction values. Further work could explore a different model that would not permit negative values. e.g. gamma GLMs [24].

A useful next step is to develop a model to predict the corresponding end-to-end duration of a state size instead of the manual approach used in this experiment. The end-to-end duration is system-generated and dependent on the system state size.

References

1. Asyabi, E., Wang, Y., Liagouris, J., Kalavri, V., Bestavros, A.: A new benchmark harness for systematic and robust evaluation of streaming state stores. In: Proceedings of the Seventeenth European Conference on Computer Systems (2022)

2. Zhang, F., Chen, H., Jin, H.: Simois: a scalable distributed stream join system with skewed workloads. In: 2019 IEEE 39th International Conference on Distributed Computing Systems (ICDCS) (2019)
3. Fang, J., Zhang, R., Fu, T., Zhang, Z., Zhou, A., Zhu, J.: Parallel stream processing against workload skewness and variance. In: Proceedings of the 26th International Symposium on High-Performance Parallel and Distributed Computing (2017)
4. Runsewe, O., Samaan, N.: Cloud resource scaling for big data streaming applications using a layered multi-dimensional hidden Markov model. In: 2017 17th IEEE/ACM International Symposium on Cluster, Cloud and Grid Computing (CCGRID) (2017)
5. Zhang, Q., Yang, L.T., Yan, Z., Chen, Z., Li, P.: An efficient deep learning model to predict cloud workload for industry informatics. IEEE Trans. Ind. Inform. **14**, 3170–3178 (2018)
6. Floratou, A., Agrawal, A., Graham, B., Rao, S., Ramasamy, K.: Dhalion: self-regulating stream processing in heron. Proc. VLDB Endow. **10**, 1825–1836 (2017)
7. Kalavri, V., Liagouris, J., Hoffmann, M., Dimitrova, D., Forshaw, M., Roscoe, T.: Three steps is all you need: fast, accurate, automatic scaling decisions for distributed streaming dataflows. In: 13th (USENIX) Symposium on Operating Systems Design and Implementation (OSDI 18) (2018)
8. Vogel, A., Griebler, D., Danelutto, M., Fernandes, L.G.: Self-adaptation on parallel stream processing: A systematic review. Concurrency Comput. Pract. Experience **34**, e6759 (2021)
9. Mohamed, S., Forshaw, M., Thomas, N.: Automatic generation of distributed run-time infrastructure for internet of things. In: 2017 IEEE International Conference on Software Architecture Workshops (ICSAW) (2017)
10. Mohamed, S., Forshaw, M., Thomas, N., Dinn, A.: Performance and Dependability evaluation of distributed event-based systems: a dynamic code-injection approach. In Proceedings of the 8th ACM/SPEC on International Conference on Performance Engineering (2017)
11. Rameshan, N., Liu, Y., Navarro, L., Vlassov, V.: Hubbub-scale: towards reliable elastic scaling under multi-tenancy. In: 2016 16th IEEE/ACM International Symposium on Cluster, Cloud and Grid Computing (CCGrid) (2016)
12. Karakaya, Z., Yazici, A., Alayyoub, M.: A comparison of stream processing frameworks. In: International Conference on Computer and Applications (ICCA) (2017)
13. Van Dongen, G., Van Den Poel, D.: Influencing factors in the scalability of distributed stream processing jobs. IEEE Access **9**, 109413–109431 (2021)
14. Tucker, P., et al., NEXMark–A Benchmark for Queries over Data Streams (DRAFT). 2008, Technical report, OGI School of Science & Engineering at OHSU, September
15. Tangwongsan, K., Hirzel, M., Schneider, S.: Sliding-Window Aggregation Algorithms (2019)
16. Jayasekara, S., Harwood, A., Karunasekera, S.: A utilization model for optimization of checkpoint intervals in distributed stream processing systems. Future Gener. Comput. Syst. **110**, 68–79 (2020)
17. Joseph, V.R., Vakayil, A.: SPlit: an optimal method for data splitting. Technometrics **64**, 166–176 (2021)
18. Nguyen, Q.H., et al.: Influence of data splitting on performance of machine learning models in prediction of shear strength of soil. Math. Prob. Eng. **2021**, 1–15 (2021)
19. Dobbin, K.K., Simon, R.M.: Optimally splitting cases for training and testing high dimensional classifiers. BMC Med Genomics **4**, 1–8 (2011)
20. Kuhn, M., Johnson, K.: Feature engineering and selection: a practical approach for predictive models (2019)
21. Carbone, P., Ewen, S., Fóra, G., Haridi, S., Richter, S., Tzoumas, K.: State management in Apache Flink®: consistent stateful distributed stream processing. Proc. VLDB Endow. **10**, 1718–1729 (2017)
22. Littler, S.: The importance and effect of sample size. https://select-statistics.co.uk/blog/importance-effect-sample-size/. Accessed 24 Apr 2022

23. Zhang, Z., Li, W., Qing, X., Liu, X., Liu, H.: Research on optimal checkpointing-interval for flink stream processing applications. Mobile Netw. Appl. **26**, 1950–1959 (2021)
24. Mazumdar, M., et al.: Comparison of statistical and machine learning models for healthcare cost data: a simulation study motivated by Oncology Care Model (OCM) data. BMC Health Serv. Res. **20**, 350 (2020). https://doi.org/10.1186/s12913-020-05148-y

Edge-cloud Computing

Edge Performance Analysis Challenges
in Mobile Simulation Scenarios

Cristina Bernad[1], Pedro J. Roig[1], Salvador Alcaraz[1], Katja Gilly[1(✉)],
and Sonja Filiposka[2]

[1] Department of Computer Engineering, Miguel Hernandez University, Elche, Spain
{cbernad,proig,salcaraz,katya}@umh.es
[2] Faculty of Computer Science and Engineering, Ss. Cyril and Methodius University,
Skopje, North Macedonia
sonja.filiposka@finki.ukim.mk

Abstract. Multi-Access Edge Computing (MEC) based services are
becoming very popular in research and innovation areas as there is a high
expectation to solve many automation and security problems through
wireless connections gathering streaming data that is processed at the
edge or cloud layer of the network. Research efforts in this direction
normally either stay at the theoretical level, or the heuristics are implemented on simulators that mainly cover an isolated part of the network
architecture as experimenting real end-to-end scenarios implies the use
of expensive infrastructure that is not normally available in research
centres. This paper deals with a simulation framework developed for
analysing MEC resource allocation algorithms performance covering the
access network, edge and cloud infrastructure and the challenges we
found during the process.

Keywords: Edge and cloud computing · end-to-end simulation ·
Performance analysis

1 Introduction

There is a huge interest nowadays in the benefits and features that will be provided by the applications and services based on the Multi-Access Edge Computing (MEC) paradigm [1]. A variety of services including tactile Internet, on skin
computing, extended reality, industry 4.0 and autonomous driving are expected
to be deployed on top of computing and network infrastructures located at the
edge of the network close to the end users. The main advantage of this approach
is to enable the use of latency sensitive services that require high computing
power and work in real or near real-time. These edge services require latencies of
less than 1ms as perceived by end users, corresponding to the ultra low latency
first promised by 5G and now expected by 6G [2].

To enable efficient use of resources and to provide the required performances
for a multitude of edge services, the deployment of extra computing infrastructure in conjunction with the network access points (base stations) must

K. Gilly and N. Thomas (Eds.): EPEW 2022, LNCS 13659, pp. 151–166, 2023.
https://doi.org/10.1007/978-3-031-25049-1_10

be accompanied with a management architecture that will be able to successfully orchestrate all processes related to the lifecycle of edge services. Unlike cloud computing, where the hosting infrastructure is located in a large datacentre, in edge computing the service provider needs to manage a number of micro datacentres, each providing edge services to the users within the service area relative to the co-located network access point [3]. Each micro datacentre is fitted with a small number of MEC hosts that provide the virtual computing infrastructure capable of hosting edge services. Thus, it is evident that the notion of geographical closeness of the user to the edge computing resources is the main characteristics of the MEC architecture that enables the implementation of ultra low latency edge services.

However, due to the mobility of end-users, MEC orchestration does not involve only efficient resource provisioning, making sure that edge services are spawned on the nearest possible MEC host thus yielding to the best quality of service (QoS) and experience (QoE) [4]. It is even more paramount that once instantiated, the edge service performances are continuously closely monitored, so that the system can intelligently respond to the changing environment and maintain the high QoS at all times. For high mobility urban deployments this means that the orchestrator must progressively migrate the edge service to other MEC hosts as the mobile user leaves one service area and enters another one which is served by a different network access point. In effect, this means that the MEC system will need to implement the so-called "follow me" behaviour [5]. Given that the MEC hosts resources are scarce compared to a full blown cloud datacentre, it is imperative that all MEC resource management techniques employed are very efficient and carefully configured.

Furthermore, the perceived performances of the MEC services are tightly coupled with the performances of the underlying network access infrastructure that is used to access them. This implies that the MEC system can not be designed and evaluated separately. When analysing the performances of different approaches regarding MEC orchestration and resource management, one must always consider the symbiotic relationship between the MEC hosts and the access network. Only in this case can the evaluation be done end-to-end, i.e. from the user perspective, which is necessary to ensure that the QoE corresponds to the QoS supplied by the MEC system.

The purpose of this paper is to investigate the impact of the performances of the access network to the overall edge service performances as perceived by the end-users. We aim to analyse the end-to-end latency experienced when using edge services in an urban highly mobile scenario, such as using edge services to augment smart vehicle capabilities. The main research question investigated is how much influence does the access network performance have over the end-to-end delay for edge services. For these purposes we have conducted a series of simulations using an urban MEC enabled 5G mobile scenario. Our results show that the access network delay can play a significant part in the overall end-to-end delay of edge services, especially in scenarios where there is no uniform distribution of users per base station.

In particular, the contributions of this paper include:

- definition of a simulation environment combining multiple simulation tools that are used to define different parts of the overall 5G MEC infrastructure
- analysis of the end-to-end latency when a mobile user uses a MEC service in a 5G urban environment, as a sum of the access network and core MEC network delays,
- identification of the impact of the variable access network delay on the overall end-to-end delay under widely ranging user loads.

The structure of the rest of this paper is as follows: in the next section we discuss the related work on the topic of MEC orchestration and performance analysis giving an overview of the proposed techniques for resource management and how their performances are investigated. In Sect. 3 we describe the simulation environment that we built in order to create a complete scenario that can track the end-to-end delay of edge services. Section 4 focuses on the obtained results from our simulation environment and here we analyse the impact of different characteristics on the MEC performances in terms of end-to-end delay. Finally, the last section concludes the paper and discusses future research topics.

2   Related Work

For the successful implementation of MEC architectures as a sustainable ecosystem, a holistic integration of the traditional telecommunications and IT virtualisation systems is needed [6]. By providing computing infrastructure at the edge of the radio network MEC promises reduced latency to the mobile end users. For these purposes an intelligent MEC orchestration is needed that will coordinate a network of distributed MEC platforms (micro datacentres) each offering MEC services in the corresponding radio service area. Taking into consideration the proliferation of IoT devices that are expected to be connected to the 5G MEC ecosystem [7], the orchestration solution must support a very large number of end users and a wide variety of services. From this perspective MEC service orchestration necessary for integrating edge computing and 5G networks needs to be seen as a holistic endeavour many technical challenges and optimisation opportunities still open for research [8].

One of the main aspects of MEC service orchestration is efficient resource management [9]. The dynamic nature of the mobile users, combined with the severely constrained resources and heterogeneous QoS policies makes MEC resource management far more challenging when compared to cloud computing. There have been many architectures, infrastructure solutions and resource management algorithms that have been proposed to tackle this problem [10]. From an optimisation point of view some of the main challenges that are being addresses include the problem of resource allocation, or the development of techniques to optimise the resource utilisation by identifying the most suitable computing resource where a newly requested MEC service will be initialised, but also resource migration, which focuses on the development of algorithms that will migrate or move to a different, more suitable, MEC host an already active MEC service in order to ensure continuously high QoS even when the user frequently changes base stations and thus increases the distance to the serving MEC host. When it comes to resource allocation algorithms there are many approaches proposed by researchers based on greedy approaches [11], game theory [12] or deep learning [13]. When it comes to resource migration, the current body of research is much smaller, with the interest in implementing dynamic resource migration rising in the past few years [14]. The two main groups of approaches are reactive, where the migration process is activated after an initial decrease in performances has been detected such as in [15], and proactive, that use different techniques to predict user mobility and thus preemptively migrate the service in order to minimise migration latency as is the case in [16]. Lately, the state of the art research is moving towards incorporating intelligence into the MEC service orchestration activities that is the flagship characteristic of 6G [17].

To analyse the performances of the vast variety of different approaches to optimising MEC service orchestration and, in particular, resource management, many researchers are using simulation frameworks [18]. For the purposes of investigating the initial behaviour of resource management algorithms in many cases extensions of the CloudSim simulators are used [19] based on discrete event simulation. This simulator however lacks the ability to simulate the radio access network and the end-users of the MEC services. When researchers would like to focus on the radio access network part and the performances of the communication between the user and the service provider infrastructure then the most prominently used simulator becomes OMNeT++, an open source tool that comes with a wide span of network characteristics that can be simulated using discrete event simulation [19]. However, the extensions of OMNeT++ that include the notion of edge computing MEC hosts such as FogNetSim++ [20] do not support resource migration.

Thus, there are still many open challenges when it comes to the holistic simulation of 5G enabled MEC infrastructure that will incorporate all aspects of service orchestration from the application layer, to the networking layer, infrastructure and mobility. Due to the complexity of the ecosystem, it is common to investigate the performances of a proposed component or an algorithm in a small scale scenario that can be more easily defined and that will not take a lot of time to execute and then analyse. For example, many evaluations are done with a very small number of devices: 15 placed on one parallel road as in [21] or 30 placed in a square 2 by 2 km area moving randomly as in [22]. While these setups can serve a very good purpose to validate the model, real life MEC scenarios are expected to run on a much larger scale, thus requiring effort and time to investigate the performances, problems and issues that might occur in dense, large scale scenarios.

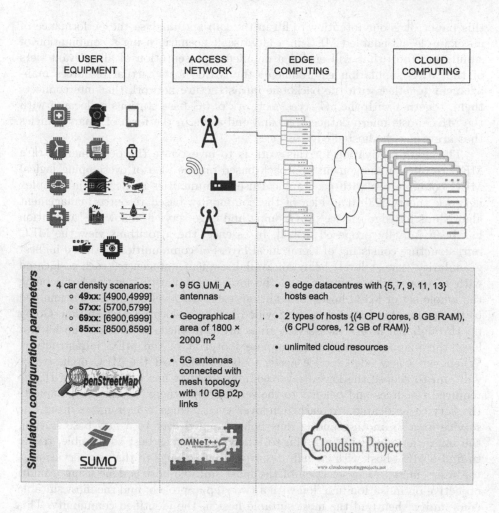

Fig. 1. Simulation configuration parameters organised per layers

3   5G Enabled MEC Simulation Environment

One of the main approaches to investigating the performances of a newly proposed MEC management technique or component is to use tools that will enable the creation of a simulation scenario that will incorporate a stochastic representation of a real life situation, such as a large scale urban environment with MEC service powered vehicles. This enables researchers to analyse how different parameters of the simulation scenario influence the overall performances of the system. As already discussed in the related work section, it is, however, more common to analyse MEC performances in a relatively small scale system (one crossroad, simple highway example or one two-way street, etc.) due to the high computing requirements of setting up a more complex simulation scenario. With this paper, it is our intention to fill in the gap and analyse the performance of an example 5G enabled MEC in a large scale scenario using a combination of simulation tools that will enable a detailed representation of all relevant tiers of the MEC simulation environment: the MEC hosts virtual resources management together with the backbone infrastructure network that interconnects them, together with the 5G access network of 5G base stations co-located with the MEC hosts micro datacentres, and end-users in the form of smart vehicles that are using the hosted edge services.

The main motivation for this work is to investigate the performances of a MEC resource management approach based on the idea of overlapping logical MEC communities with geographical user communities as defined in complex network theory [23]. The idea of the community based resource management algorithms for edge service provisioning and edge service "follow-me" migration has been initially proposed in [24]. In essence, the algorithms view the MEC infrastructure consisting of hierarchical layers of communities, with the highest granularity (lowest level) being all available micro datacentres each co-located with a separate base station, and the lowest granularity (highest level) being the whole set of MEC hosts. In addition, each low level community is mapped to the geographical service area provided by the co-located base station. Going up through the layers the service areas are combined, so that on the highest level the mapped area is the complete footprint of the MEC infrastructure. Whenever a new edge service needs to be provisioned the MEC orchestrator will aim to choose the lowest layer community that has a MEC host with the required resources and belongs to the service area where the user that requests the service is located. At each handover event, when a user moves from one service area to another and is thus being handed over to a new base station, the migration algorithm will aim to follow the user as best as possible, trying to find a MEC host with enough resources that belongs to the new service area the user entered. The selection of the most suitable host is done using a multi objective optimisation function with a two step approach: find the most suitable community, then find the most suitable host in the identified community. This allows for different types of optimisations to be implemented on the community level, such as load balancing or consolidation of the virtual resources of MEC hosts that belong to the same micro datacentre.

The performances of the community based resource management approach have been analysed in [25] and it has been shown that it greatly outperforms other typical approaches to MEC resource management such as the placement and migration policies readily available in CloudSim including first-fit, load balancing, server consolidation, median average deviation and combinations thereof. These promising results have led to defining a proactive version of the proposed approach, that uses machine learning to predict the user mobility and hence handover events initiating the migration process beforehand and thus reducing the edge service latency introduced by the migration process [26]. It should be noted that when employing the model proactively, there is always a chance of wrong prediction in the form of false migration or false non-migration that can reduce the benefits of the follow-me behaviour depending on the reliability of the predictive model. The cost that needs to be paid in terms of increased latency and lost capacity is discussed in details in [26]. However, the simulation environment that has been used to analyse the performances of the community based resource management approaches has been focusing on the MEC infrastructure and the mobility of the end-users, not taking into account the access network in between. In other words, the simulations performed have presumed that the access network has limitless capacity and that the service areas of the dummy base stations are roughly the same size calculated based on a Voronoi diagram using Thiessen polygons [27] incurring no possibility to analyse the impact of the network access variability on the overall system.

Given that the proposed resource management algorithms have shown high potential for use in the simplified simulation scenarios, the purpose of this paper is to go one step further and analyse the performance of the approach by focusing on the end-to-end delay from the user to the edge service by creating a full blown simulation scenario that will also include a more realistic model of the 5G access network. This analysis will allow us to investigate if the access network infrastructure imposes some limitations that must be taken into account when designing the logic of MEC orchestrators.

For these purposes we have extended the initial simulation environment and created a workflow that brings together several independent simulation tools to best describe the different aspects of the simulation scenario. For the purposes of simulating the MEC infrastructure we have chosen CloudSim [28] adapted for MEC as it is one of the most popular simulation tools that can be used to define all MEC hosts and their specific virtual resources, and specify the backbone network that interconnects them including network devices such as switches and routers. CloudSim supports a range of different policies when it comes to defining how the virtual resources are used by the instantiated virtual machines, and it can be easily extended with additional placement and migration management policies which was done in order to implement the mechanisms of the community based placement and migration algorithms. One of the main characteristics of CloudSim is that it supports large scale scenarios that can potentially involve hundreds of hosts and thousands of virtual machines. To be able to adequately simulate the behaviour of the 5G access network we have chosen the OMNeT++ [29] simulator that has been extended with a specialised 5G module [30]. OMNeT++ is one of the top choices when it comes to simulation of access and core networks and it enables definitions of highly detailed large scale scenarios that incorporate a wide range of tunable parameters and statistical models such as radio wave propagation, fading, channel interference, etc. The 5G extension module builds on the built-in 4G module so that all newly introduced specifics on the physical and logical layers of 5G networks are added and available for use in the OMNeT++ simulation scenarios. Finally, to be able to simulate a large scale urban mobile scenario, the SUMO [31] mobility simulator has been used. SUMO enables the user to specify a real life area described using OpenStreetMaps wherein the generated vehicles are randomly moving while obeying all traffic rules and restrictions as they are defined in the OpenStreetMaps metadata files. Using SUMO one can generate the realistic mobility pattern of thousands of vehicles in a given urban area. By varying the number of vehicles different scenarios can be constructed: from light dispersed traffic to heavily congested traffic. Figure 1 summarises the configurations that we have considered at the different simulation levels in a general IoT-MEC scenario.

The simulation workflow has been defined such that initially SUMO creates an example mobility pattern. Based on the entering and leaving events of each vehicle in the SUMO generated simulation the start and end time of edge service instances are defined so that each vehicle is served by a separate MEC service. By considering a single MEC service per vehicle (corresponding to the use of an inbuilt infotainment system) we can create a simulation scenario that enables a one by one mapping of vehicles and MEC service users, making it easier to interpret the obtained results. The SUMO generated mobility pattern is then introduced in OMNeT++ that simulates the continuous communication between each user (vehicle) and its corresponding edge service that is located behind the base stations. For the purposes of simulating the access network, the MEC hosts in OMNeT++ are modelled as one dummy host directly connected to all base stations on which all MEC services are instantiated when a new vehicle enters the simulation and destroyed when the vehicle leaves the simulation. The log output from OMNeT++ is analysed to identify to handover events, that are then fed with the rest of the information to CloudSim that manages the placement and migrations of each edge service in the form of a virtual machine. In CloudSim the MEC hosts infrastructure is represented in all details, with the corresponding number of hosts per base station, and the optical network that interconnects the base stations host clusters. The vehicle entry and exit event from SUMO define the start and end time of each MEC service, while the base station defines the user service area and is used by the placement and migration algorithms. The logs from all three simulators are combined together so that the delays in the different parts of the network are identified (delay in the 5G access network and delay in the MEC backbone infrastructure network). The separation of duties: SUMO - mobility, OMNeT++ - access network, CloudSim - MEC computing enables the use of the tools in the described workflow without any concurrent execution. The access network delay and the MEC core network delay are combined by using the base stations as points of reference, since each base station is directly connected to its own cluster of MEC hosts. These two delays are then combined together to form the end-to-end delay as perceived by each user (vehicle) that communicates with its corresponding edge service. A number of pre and post processing scripts have been developed to be able to use the output of one simulation tool as input for the next one, as well as to synchronise the outputs of the simulators so that the end-to-end latency can be obtained. The detailed description of the proposed workflow and the related pre and post processing scripts together with a step by step example are made available on GitHub [33] under the Creative Commons licence. Following the

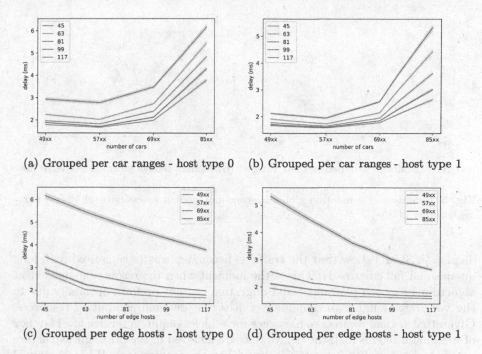

(a) Grouped per car ranges - host type 0   (b) Grouped per car ranges - host type 1

(c) Grouped per edge hosts - host type 0   (d) Grouped per edge hosts - host type 1

Fig. 2. Evolution of average edge network delays with their 95% confidence intervals when varying number of cars and edge hosts allocated at the MEC layer

workflow, one can create a vast variety of simulation scenarios using different options such as other placement and migration algorithms, multiple services per vehicle, different locations and network infrastructure setup, etc.

4   Results and Discussion

The created output log files from CloudSim and OMNeT++ have been analysed separately and combined in order to be able to determine the delays in the radio access network and the delays in the backbone optical infrastructure that connects the MEC hosts and the base stations. For each communication event the two types of delays have been synchronised and added together to determine the user experienced end-to-end delay when accessing a MEC service.

In Fig. 2 the average delay in ms that occurs in the MEC backbone network is presented for different number of vehicles, or traffic densities, when the MEC infrastructure is built using different number of MEC hosts and hosts types. As it can be expected, the average delay increases as the traffic in the MEC service area intensifies and is very much influenced by the number of MEC hosts that are available at each base station. The influence of the host type (small - 0 vs large - 1) is more pronounced for smaller number of nodes. The steep increase in delay is experienced for the 85xx car scenarios, wherein the log output and graphical

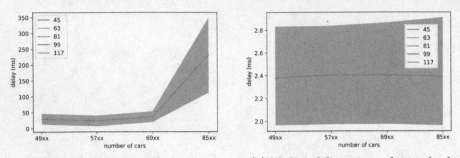

(a) Average delays grouped per car ranges   (b) Median delays grouped per edge hosts

Fig. 3. Evolution of simulation global average and median access network delays varying number of cars

display in SUMO show that the traffic is becoming heavily congested with long queues and full crossroads. This is the moment when the resource management algorithms for both placement and migration can not perform optimally due to the very large number of dynamic services that demand computing resources. Given that normal awareness messages for vehicles require a service level latency of maximum 100 ms [32], the delays that are incurred in the MEC infrastructure network shown in Fig. 2 are much lower than the requirement. When combined with the promised latency of 1 ms that should be provided by the 5G network, the corresponding end-to-end delay should also be well within the requirements.

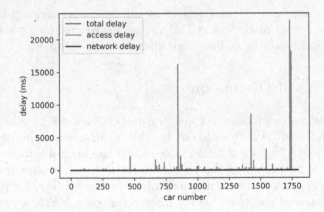

Fig. 4. Access network connection delays for the 4928 vehicles simulation

When analysing the the overlapping values for all edge datacentre configuracion of the radio access network delay obtained from the OMNeT++ simulator presented in Fig. 3a, it can be seen that the obtained average delay and its 95% confidence intervals in the access network is much higher than the "expected" 5G performances. For the scenarios from light up to moderate traffic (49xx to

69xx vehicles) the communication delay between the end users and the base stations is somewhat steady around 30 ms, but becomes much higher and more variable for the congested traffic scenario of 85xx services. In fact, we can compare the average values with the median and 25th and 75th percentile values that are represented in Fig. 3b observing that median values are much smaller than average values and this can only happen if there is a high variation in the simulated access network delays as median values are about 20 times smaller than the average at the least congested scenarios. This assumption is confirmed observing Fig. 4 that shows how the average of access delays per vehicle during the simulation time of one of the 49xx simulations gets very high compared to network delays that are below 2 ms (see Fig. 3) at certain moments because of some kind of connectivity problems that we will analyse below.

The average delay in the radio access network has been combined with the delay in the MEC backbone network to present the end-to-end average network delay representing the total communication delay from the end-user to the corresponding MEC service in Fig. 5. It is evident that the end-to-end delay is dominated by the radio access network delay, and any more significant difference in the delays for the different scenarios is visible only for the heavily congested traffic scenario with the maximum number of MEC services when the MEC infrastructure was also having issues keeping track of users and optimally migrating their MEC services. The lower two images that show the end-to-end delay as a function of the number of edge hosts for different host types Fig. 5c and Fig. 5d provide a more clear view of the differences for different loads (number of MEC services), demonstrating the large variability in performances for the heaviest scenario where the delay varies from 70 to 350 ms. The obtained results for the end-to-end average delay have not been expected by the authors when comparing the achieved performances with the specifications and requirements for the 5G radio networks. However, the results clearly show that one can not rely on trusting that another component of an integrated system (the 5G-MEC in this case) will perform as expected in all scenarios, especially as large scale and dynamic as the ones presented here. Thus, the results showcase the need to testing the performances of the proposed solutions in large-scale urban, real-life settings so that one can analyse what are the requisites for the optimal system performances.

Aiming to delve further into the results and understand why the 5G access network is not performing as well as expected, we have conducted a series of more detailed analysis of the output logs obtained from OMNeT++. It is interesting to note that the logs show that as the simulation time progresses an increasing amount of users are all being connected to a single base station (no. 2) which puts a tremendous strain on the particular base station and incurs much higher delays for the users (please see Fig. 6). This kind of behaviour has not been observed in the initial performance analysis [25] since the concept of Voronoi cells used there ensured a relatively uniform distribution of users per base station service area. However, it turns out that when incorporating the propagation and fading models available in OMNeT++ the service areas of the base stations change

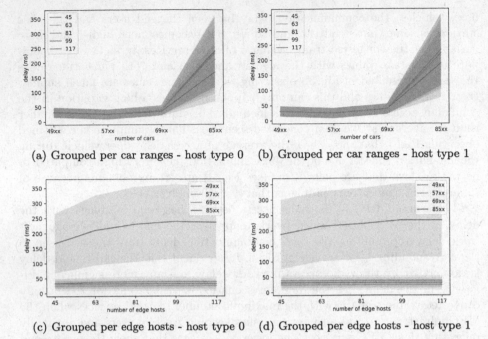

(a) Grouped per car ranges - host type 0   (b) Grouped per car ranges - host type 1

(c) Grouped per edge hosts - host type 0   (d) Grouped per edge hosts - host type 1

Fig. 5. Evolution of average end-to-end network delays varying number of cars and edge hosts allocated at the MEC layer

more dramatically making the service area of base station 2 much larger than expected. Since this base station serves one of the busiest area of the city centre chosen for the simulation, the result is the heavy user generated demand on this base station that greatly affects the overall end-to-end performances.

In essence this performance analysis presents two very important challenges when studying the design and implementation of MEC systems. First, that the performances of such a system must be analysed holistically taking into account all communication layers and all components on the path between the user and the MEC service. And second, that the optimal (or sub-optimal) design of one system (5G in this case) can greatly influence the behaviour of the other (MEC). What was initially taken as a good radio access design with base stations distribution to cover similarly sized service area cells has proven to be a poor choice when incorporating other aspects of the radio network. Unfortunately the choice of where to locate the base stations when setting the parameters of the OMNeT++ simulation scenario is somewhat tedious since there is no service area visualisation available for the 5G module yet.

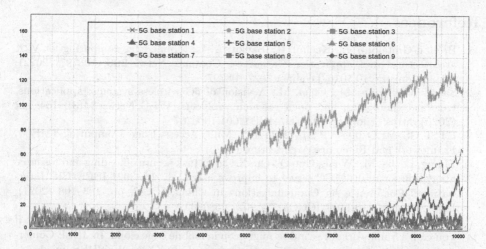

Fig. 6. OMNeT++ figure of the evolution of user equipment connections to 5G base stations during simulation time of the 4951 vehicles simulation

5 Conclusion

Discrete event simulation tools are a very popular choice for performance analysis of service orchestration in edge computing systems. They enable researchers define scenarios that will be used for testing the behaviour of different MEC components using random generators of service demand and user mobility and implementing different stochastic models that aim to reproduce realistic events in the simulation environment. With MEC becoming one of the most popular service offer for 5G networks, the research on optimal integration of this two systems into one ecosystem must take on a holistic approach creating a simulation environment that will bring together all aspects of both MEC and 5G.

The simulation based performance analysis outlined in this paper presents a workflow that combines the most popular simulation tools for analysing MEC and 5G that can be used to create an all encompassing simulation scenario that will incorporate all relevant layers starting from the application where MEC service instances reside, to the network layer where the MEC infrastructure is connected to the 5G backbone and the radio access network connects the users in the ecosystem, to the physical layer with the dynamic nature of radio communication and the mobility of the end-users. Our results show that to achieve overall optimal end-to-end latency, one must carefully design and analyse all characteristics of both MEC and 5G, and using an oversimplified model for one of the systems might yield to unexpected challenges that will need to be addressed at a later stage. The decisions on the design of the MEC and 5G infrastructures must be taken holistically acknowledging that certain aspects such as base stations with increased service areas need to reflect in the MEC infrastructure design by, for an example distributing more computing power to the base stations that are expected to serve an increased number of users.

References

1. Bréhon-Grataloup, L., Kacimi, R., Beylot, A.-L.: Mobile edge computing for V2X architectures and applications: a survey. Comput. Netw. **206**, 108797 (2022). https://doi.org/10.1016/j.comnet.2022.108797
2. Saad, W., Bennis, M., Chen, M.: A vision of 6G wireless systems: applications, trends, technologies, and open research problems. IEEE Netw. **34**(3), 134–142 (2020). https://doi.org/10.1109/MNET.001.1900287
3. ETSI GS MEC 003 V2.2.1 (2020–12): Multi-Access Edge Computing (MEC); Framework and Reference Architecture (2020)
4. Zhang, L., Jia, M., Wu J., Guo Q., Gu, X.: Joint task secure offloading and resource allocation for multi-MEC server to improve user QoE. In: 2021 IEEE/CIC International Conference on Communications in China, ICCC, pp. 103–108 (2021). https://doi.org/10.1109/ICCC52777.2021.9580302
5. Doan T.V., Fan Z., Nguyen G.T., Salah H., You D., Fitzek, F.H.P.: Follow me, if you can: a framework for seamless migration in mobile edge cloud. In: IEEE Conference on Computer Communications Workshops, INFOCOM WKSHPS, pp. 1178–1183 (2020). https://doi.org/10.1109/INFOCOMWKSHPS50562.2020.9162992
6. Taleb, T., Samdanis, K., Mada, B., Flinck, H., Dutta, S., Sabella, D.: On multi-access edge computing: a survey of the emerging 5G network edge cloud architecture and orchestration. IEEE Commun. Surv. Tutor. **19**(3), 1657–1681 (2017). https://doi.org/10.1109/COMST.2017.2705720
7. Alam, M., Rufino, J., Ferreira, J., Ahmed, S.H., Shah, N., Chen, Y.: Orchestration of microservices for iot using docker and edge computing. IEEE Commun. Mag. **56**(9), 118–123 (2018). https://doi.org/10.1109/MCOM.2018.1701233
8. Guo, Y., Qiang D., Wang, S.: Service orchestration for integrating edge computing and 5G network: state of the art and challenges. In: 2020 IEEE World Congress on Services (SERVICES), pp. 55–60. IEEE (2020). https://doi.org/10.1109/SERVICES48979.2020.00026
9. Hong, Ch., Varghese, B.: Resource management in fog/edge computing: a survey on architectures, infrastructure, and algorithms. ACM Comput. Surv. **52**(5), 1–37 (2019). https://doi.org/10.1145/3326066
10. Mijuskovic, A., Chiumento, A., Bemthuis, R., Aldea, A., Havinga, P.: Resource management techniques for cloud/fog and edge computing: an evaluation framework and classification. Sensors **21**(5), 1832 (2021). https://doi.org/10.3390/s21051832
11. Fan, Y., Wang, L., Wu, W., Du, D.: Cloud/edge computing resource allocation and pricing for mobile blockchain: an iterative greedy and search approach. IEEE Trans. Comput. Soc. Syst. **8**(2), 451–463 (2021). https://doi.org/10.1109/TCSS.2021.3049152
12. Roostaei, R., Dabiri, Z., Movahedi, Z.: A game-theoretic joint optimal pricing and resource allocation for mobile edge computing in NOMA-based 5G networks and beyond. Comput. Netw. **198**, 108352 (2021). https://doi.org/10.1016/j.comnet.2021.108352
13. Dong, R., She, Ch., Hardjawana, W., Li, Y., Vucetic, B.: Deep learning for hybrid 5G services in mobile edge computing systems: learn from a digital twin. IEEE Trans. Wirel. Commun. **18**(10), 4692–4707 (2019). https://doi.org/10.1109/TWC.2019.2927312
14. Wang, S., Xu, J., Zhang, N., Liu, Y.: A survey on service migration in mobile edge computing. IEEE Access **6**, 23511–23528 (2018). https://doi.org/10.1109/ACCESS.2018.2828102

15. Wu, Ch., Peng, Q., Xia, Y., Ma, Y., Zheng, W., Xie, H., et al.: Online user allocation in mobile edge computing environments: a decentralized reactive approach. J. Syst. Archit. **113**, 101904 (2021). https://doi.org/10.1016/j.sysarc.2020.101904
16. Slamnik-Krijetorac, N., Carvalho de Resende, H.C., Donato, C., Latr, S., Riggio, R., Marquez-Barja, J.: Leveraging mobile edge computing to improve vehicular communications. In: 2020 IEEE 17th Annual Consumer Communications & Networking Conference (CCNC), pp. 1–4. IEEE (2020). https://doi.org/10.1109/CCNC46108.2020.9045698
17. Al-Ansi, A., Al-Ansi, A.M., Muthanna, A., Elgendy, I.A., Koucheryavy, A.: Survey on intelligence edge computing in 6G: characteristics, challenges, potential use cases, and market drivers. Future Internet **13**(5), 118 (2021). https://doi.org/10.3390/fi13050118
18. Svorobej, S., Takako Endo, P., Bendechache, M., Filelis-Papadopoulos, C., Giannoutakis, K.M., Gravvanis, G.A., et al.: Simulating fog and edge computing scenarios: an overview and research challenges. Future Internet **11**(3), 55 (2019). https://doi.org/10.3390/fi11030055
19. Bendechache, M., Svorobej, S., Takako Endo, P., Lynn, T.: Simulating resource management across the cloud-to-thing continuum: a survey and future directions. Future Internet **12**(6), 95 (2020). https://doi.org/10.3390/fi12060095
20. Qayyum, T., Malik, A.W., Khattak, M.A.K., Khalid, O., Khan, S.U.: FogNetSim++: a toolkit for modeling and simulation of distributed fog environment. IEEE Access **6**, 63570–63583 (2018). https://doi.org/10.1109/ACCESS.2018.2877696
21. Tang, W., Zhao, X., Rafique, W., Qi, L., Dou, W., Ni, Q.: An offloading method using decentralized P2P-enabled mobile edge servers in edge computing. J. Syst. Archit. **94**, 1–13 (2019). https://doi.org/10.1016/j.sysarc.2019.02.001
22. Feng, J., Yu, F.R., Pei, Q., Chu, X., Du, J., Zhu, Li.: Cooperative computation offloading and resource allocation for blockchain-enabled mobile-edge computing: a deep reinforcement learning approach. IEEE Internet Things J. **7**(7), 6214–6228 (2019). https://doi.org/10.1109/JIOT.2019.2961707
23. Filiposka, S., Juiz, C.: Community-based complex cloud data center. Phys. A: Stat. Mech. Appl. **419**, 356–372 (2015). https://doi.org/10.1016/j.physa.2014.10.017
24. Filiposka, S., Mishev, A., Gilly, K.: Community-based allocation and migration strategies for fog computing. In: 2018 IEEE Wireless Communications and Networking Conference (WCNC). https://doi.org/10.1109/WCNC.2018.8377095
25. Filiposka, S., Mishev, A., Gilly, K.: Mobile-aware dynamic resource management for edge computing. Trans. Emerg. Telecommun. Technol. **30**(6), e3626 (2019). https://doi.org/10.1002/ett.3626
26. Gilly, K., Filiposka, S., Alcaraz, S.: Predictive migration performance in vehicular edge computing environments. Appl. Sci. **11**(3), 944 (2021). https://doi.org/10.3390/app11030944
27. Abo-Zahhad, M., Sabor, N., Sasaki, S., Ahmed, S.M.: A centralized immune-voronoi deployment algorithm for coverage maximization and energy conservation in mobile wireless sensor networks. Inf. Fusion **30**, 36–51 (2016). https://doi.org/10.1016/j.inffus.2015.11.005
28. Calheiros, R.N., Ranjan, R., Beloglazov, A., De Rose, C.A.F., Buyya, R.: Cloudsim: a toolkit for modeling and simulation of cloud computing environments and evaluation of resource provisioning algorithms. Softw.: Pract. Experience **41**(1), 23–50 (2010). https://doi.org/10.1002/spe.995
29. Varga, A. Hornig, R.: An overview of the OMNeT++ simulation environment. In: 1st International Conference on Simulation Tools and Techniques for Communi-

cations, Networks and Systems Workshops (Simutools), pp. 1–10 (2008). https://doi.org/10.5555/1416222.1416290

30. Deinlein, T., German, R, Djanatliev, A.: 5G-Sim-V2I/N: towards a simulation framework for the evaluation of 5G V2I/V2N use cases. In: 2020 European Conference on Networks and Communications (EuCNC) (2020). https://doi.org/10.1109/EuCNC48522.2020.9200949

31. Alvarez-Lopez, P., Behrisch, M., Bieker-Walz, L., Erdmann, J., Flötteröd, J.P., Hilbrich, R. et al.: Microscopic traffic simulation using SUMO. In: 2018 21st International Conference on Intelligent Transportation Systems (ITSC) (2018). https://doi.org/10.1109/ITSC.2018.8569938

32. Cinque, E., Valentini, F., Persia, A., Chiocchio, S., Santucci, F., Pratesi, M.: V2X communication technologies and service requirements for connected and autonomous driving. In: 2020 AEIT International Conference of Electrical and Electronic Technologies for Automotive (AEIT AUTOMOTIVE), pp. 1–6. IEEE (2020). https://doi.org/10.23919/AEITAUTOMOTIVE50086.2020.9307388

33. Edge simulation github repository. mobile edge computing simulation in 5G environment (2022). https://github.com/EdgeSimulation. Accessed 1 July 2022

A Deterministic Model to Predict Execution Time of Spark Applications

Hina Tariq and Olivia Das[✉]

Electrical, Computer and Biomedical Engineering, Ryerson University,
Toronto, Canada
odas@ee.ryerson.ca

Abstract. This work proposes a graph-based, deterministic analytical model that predicts the execution time of spark applications. It conceptualizes the structure of the spark application as a monolithic Directed Acyclic Graph (DAG) of stages capturing the precedence relationship among all the stages of the application. The model processes every stage of the DAG using a graph traversal algorithm, combined with a fixed scheduling policy of the spark platform in context (spark platform refers to the cloud that hosts the spark cluster). We validate our model against the measured execution time obtained by running a big data query (Query-64 of TPC-DS benchmark) that involves parallel execution of a large number of stages. The query is executed on the spark cluster of Google Cloud. Our model resulted in an execution time that is at 2.85% error in comparison to the measured execution time.

Keywords: Analytical model · Deterministic model · Execution time · Spark application · Stages · Directed acyclic graph

1 Introduction

Big Data applications typically handle very large and complex data sets. Due to limitation of storage, it is a challenging task to handle such a massive amount of data in a single node. Currently, Apache Spark (henceforth referred to as spark) is an open-source, general-purpose, distributed, cluster-computing framework that enables a big data application to run in parallel on multiple nodes. Spark is general-purpose in the sense that it allows running of applications from wide variety of domains such as machine learning, graph processing and data streaming; it is considered a distributed framework since a spark application can run on multiple nodes; it is also a cluster-computing entity since a cluster of distributed nodes can be dedicated to run the same spark application. It extends the previously popular MapReduce framework to support plethora of computation types—one that includes computation of interactive queries—with higher efficiency [6]. The spark framework provides a large set of configuration parameters that influence the performance of a spark application; it is in fact a challenging proposition to figure out how alteration of these parameters can influence the

K. Gilly and N. Thomas (Eds.): EPEW 2022, LNCS 13659, pp. 167–181, 2023.
https://doi.org/10.1007/978-3-031-25049-1_11

application execution time—a critical metric of application performance. This is because owing to multitude of such parameters, it is a very time-consuming and laborious task to evaluate—through actual measurements—which parameters would take a greater effect on the execution time of a spark application [8]. Performance models come to rescue in this case. Such models are increasingly being adopted for prediction since they enable one to quickly estimate how application execution time is likely to be affected by the spark configuration parameters of interest [4].

Recently, few works have used machine learning models for predicting performance of spark applications [5,7,11]. These models require substantial amount of previously measured data to make effective predictions. Collecting such vast amount of data is a time-consuming and resource-intensive task that often comes at an exorbitant financial cost; it is time-consuming since one has to run a spark application repeatedly while varying the values of several different parameters and record the measurements for every run; it is resource-intensive since one has to acquire multiple nodes for substantially long period of time to run the experiments; and, this whole process of experimentation entails huge financial expenditure incurred to pay for the acquired cloud resources and the human operators who carry out such experiments.

Analytical models, on the other hand, can provide insight into the performance of a spark application in less time. Building such models requires substantially fewer runs of data-collection experiments (to collect measurement data for model parameter estimation) in comparison to the machine learning models. The work by Shah et al. [9] had proposed such a model that was deterministic. Their model *groups* those very stages inside a spark job that have no dependencies on each other. It *imposes* that all the stages in a group *must* finish before a succeeding dependent stage can start. This is a strong assumption because, it is possible that a succeeding stage may not be dependent on all the stages of the preceding group; yet, in this case, the succeeding stage will not be able to start even if the stages it depends upon have finished, and executor cores are available for execution. Our approach overcomes this limitation.

In this work, we propose a graph-based, deterministic analytical model that predicts the execution time of spark applications as follows: (i) We are aware that a spark application is often conceptualized hierarchically—the application consists of *jobs*, and each job contains a Directed Acyclic Graph (DAG) of *stages*. For our modeling purposes, we remove the concept of jobs from this hierarchy; instead, we conceptualize the structure of the spark application as a *flat*, monolithic DAG of stages that *directly* captures the precedence relationship among the stages across all the jobs of the application. (ii) our model processes every stage of the DAG using a graph traversal algorithm, combined with a fixed scheduling policy of the spark platform in context (spark platform refers to the cloud that hosts the spark cluster). (iii) it allocates free executors to those succeeding stages that are ready to start rather than waiting for the preceding parallel stage, with the largest execution time, to finish. To estimate the values of the input parameters of our model, we use two reference cases, one with smaller input data size, and the other with bigger input data size. We use these two cases to fetch the DAG of the spark application, and extract the values of the model input parameters such as average task execution time for a given stage, number of tasks for a given stage etc. from the spark history server. We run each reference case four times and compute the grand average of the task execution times (for each stage of the DAG) obtained from the four runs, for each case.

We walk through our model solution algorithm on a synthetic DAG. Next, we validate our model against the measured execution time obtained by running a big data query—Query-64 of TPC-DS benchmark (a popular big data benchmark [10])—on the spark cluster of Google Cloud. Query-64 is a complex query that involves parallel execution of a large number of stages. Our model results in 2.85% error in comparison to the measured execution time of the query.

The rest of the paper is organized as follows. Section 2 explains the Spark Architecture in the context of this work. Section 3 describes our proposed model. In Sect. 4, we present the results and analyse them. Section 5 enumerates some related work. Section 6 provides a brief discussion. Finally, Sect. 7 concludes the paper with suggestion on future work.

2   Apache Spark Architecture

Apache Spark facilitates parallel processing of big volume of data on a cluster of nodes [12]. Its architecture is illustrated in Fig. 1. The architecture consists of a driver (containing the spark application logic), cluster manager, and one or more executors with a specified number of cores (i.e. physical threads) per executor. These cores enable parallel computation. The driver is hosted in the master node of the cluster, and the executors are hosted in the worker nodes of the cluster (the dotted line between the two worker nodes in the figure represents the fact that there might be several such nodes). There is a SparkContext object in the driver. The SparkContext negotiates with the cluster manager to allocate the executors on the worker nodes for a specific spark application (interaction indicated using dashed lines with arrowheads in the figure). Once the executors are allocated, the driver bypasses the cluster manager and directly starts communicating with the executors (interaction indicated using solid lines in the figure).

A spark application consists of one or more jobs. The jobs can either run in sequence or in parallel. Each job constitutes one or more units of execution called stages. The workflow of execution of stages form a DAG. A stage that depends on the output data generated by the preceding stages in the DAG cannot start execution until all those preceding stages have finished. The stages that have no dependencies with one another can run in parallel. A stage consists of one or more tasks. There are some stages that are dependent on input data—we call them *dynamic stages*, and others that are independent of input data—we call them *static stages*. In a dynamic stage, the number of tasks changes if we change the input data size. The number of tasks in a static stage is independent of input data and hence does not change. A task is executed by a core of an executor on a worker node.

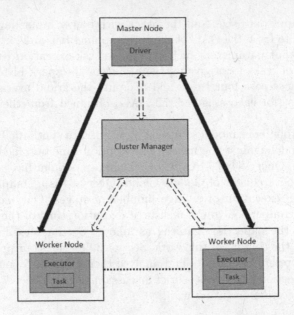

Fig. 1. Apache Spark Architecture

Figure 2 shows a synthetic spark application. It has three jobs job1, job2, and job3 in sequence. The job1 contains stage 1, job2 contains stages 2, 3, 4, 5 and job3 contains stage 6. Stages 2, 3 and 4 can run in parallel. Figure 3 shows a flat, monolithic DAG of stages of the application with the concept of job *removed* from the job-stage hierarchy. The conversion from Fig. 2 to Fig. 3 is done as follows: The output stages (the stages that do not have any succeeding stage) of a job are connected to all those stages of the next job in sequence which do not have any preceding stages. This conversion helps us in accounting for the parallel jobs (unlike [2] and [3]), in addition to the sequential jobs, of a spark application.

3   Proposed Model

In this section, first we introduce some notations. Then we describe our proposed model, and its solution algorithm. Subsequently, we describe how to estimate the model input parameters from two reference executions of a spark application on a given cloud.

3.1 Notations

S	List of stages in the DAG
T_s	Number of tasks for stage $s \in S$
e_s	Average execution time for any task of stage $s \in S$
$Succ(s)$	List of direct successors of stage $s \in S$
$Pred(s)$	List of direct predecessors of stage $s \in S$
C	List of executor cores
W	Warm-up time of the application that includes the time to start the driver and the executors
$t.ret$	returns the remaining execution time of task t
RS	Set of stages whose tasks are ready to be scheduled for execution
RT	Set of tasks ready to be scheduled for execution
ET	Set of currently executing tasks
T_{next}	Time period for the next task t to complete, $t \in ET$
T_{exec}	Execution time of the spark application
CS	Set of stages in RS that have completed (A stage is complete when all its tasks have completed execution)

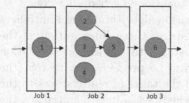

Fig. 2. A synthetic spark application with a DAG of stages for each job.

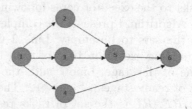

Fig. 3. The monolithic DAG of stages of the spark application of Fig. 2. Unlike Fig. 2, this DAG is flat with the concept of job *removed*.

3.2   Model Description

In this section we propose our model for prediction of execution time T_{exec} of a spark application. Our model takes as input the following parameters:

- the list of executor-cores C available to process the tasks in parallel. We assume that the total number of executor-cores $|C|$ does not exceed the total number of vCPU-cores. This is to avoid any contention in the vCPUs. Thus, the competition among the tasks will only be for the executor-cores as resource.
- the warm-up time, W of the spark application. This time includes the time needed for starting the driver and the executors on the master and the worker nodes.
- the DAG of the spark application. The DAG is expressed in terms of the list of stages S in the DAG. For each stage $s \in S$, the list of direct predecessor stages $Pred(s)$ of s and the list of direct successor stages $Succ(s)$ of s.
- For each stage $s \in S$, the number of tasks T_s.
- For each stage $s \in S$, the average task execution time e_s.

When we say "stages are ready to be scheduled for execution", we mean "the *tasks* in the given stages are ready to be scheduled for execution". While running the experiments on Google Cloud, we have observed that when multiple stages are ready to be scheduled for execution at the same time, the spark platform does a round-robin scheduling of the stages on the executor-cores. Let us take an example to explain this observation. Suppose there are three stages s1, s2, s3 that are ready to be scheduled for execution. Let s1 have 10 tasks, s2 have 5 tasks, and s3 have 2 tasks. If we have 8 executor-cores available, the spark would quite often schedule the tasks as follows: In the first round, it would schedule 1 task from each stage to each of the three executor-cores. Next, in the second round, it would do the same. However, in the third round, it would schedule 1 task each from s1 and s2 to the remaining two executor-cores; thus, three tasks of s1, three tasks of s2, and two tasks of s3 will be scheduled.

Our model essentially traverses the DAG of stages and schedules the ready tasks to the executor cores following the above mentioned scheduling policy.

Algorithm 1 presents the complete model solution algorithm. Line 1 initializes all the sets to be empty. Line 2–6: For each stage, these lines initializes the remaining execution time of each task (of the stage) to the average execution time of the stage. Line 7 adds the stages, having no predecessors, to RS (the set of ready stages). Line 8–12: These lines add the tasks of ready stages to the set RT. RT is the set of tasks ready to be scheduled. Line 13 initializes the execution time T_{exec} of the spark application to 0. Line 14 initializes T_{next} to 0. Line 15–52 processes all the stages of the DAG and schedules their tasks to the available cores. Line 16–24: These lines schedule the tasks of the ready stages in the round-robin fashion (explained earlier) to the available cores. Whichever ready task is scheduled, is removed from RT and added to the set of executing tasks ET. Line 25 computes the minimum of all the remaining execution times of the tasks in the set ET, and assigns it to T_{next}. This minimum time is the time-period after which the next executing task(s) will complete. Line 26 updates the T_{exec} by including the T_{next}. Line 27–34: Every task in ET, whose remaining execution time is greater than T_{next}, is yet to complete; otherwise the task is complete. Update the remaining execution times of the incomplete tasks (Line 29), and delete the completed tasks from ET (Line 32). Line 35–40: Identify all the stages in RS that are complete (a stage is complete whose all tasks are complete), add them to the set CS, and remove them from RS. Line 41–50: Look at every successor stage of each complete stage. If all the predecessor stages of that successor stage are complete, then add that successor stage to the set of ready stages RS and, all the tasks of that successor stage to the set of ready tasks RT. The loop of Line 15–52 repeats until the set RS is empty (Line 52), that is, until all the stages in the DAG have been processed. Line 53: Finally add the warm-up time to T_{exec}. This is the output of the algorithm that predicts the execution time of the spark application.

3.3   Estimation of Model Input Parameters

This section answers the following question: If we want to use our proposed model to predict the execution time of a spark application for an arbitrary big input data size, for example $d1$, on a given number of executor cores C of a target cloud, how can we estimate the values of the model input parameters such as $S, Succ(s), Pred(s), T_s, e_s$ for every stage $s \in S$, W?

To estimate the values of the aforementioned model input parameters, we need two reference cases, one with smaller input data size $d2$, and the other with bigger input data size $d3$. Both cases should use C executor cores of the target cloud. With respect to each case, we run the spark application four times for the case and then, for every stage of the DAG, we take the average of the task execution times obtained from those four runs per case.

Estimation of DAG for input data size $d1$: We have observed that the DAG of the spark application does not change for different input data sizes. The DAG represented by $S, Succ(s)$ and $Pred(s)$ can therefore be obtained from the spark history server of any one execution corresponding to the reference case related to either $d2$ or $d3$.

Estimation of T_s and e_s of each stage s for the input data size $d1$: Let the number of tasks for stage s for input data sizes $d1$, $d2$ and $d3$ be $T_s(d1)$, $T_s(d2)$ and $T_s(d3)$ respectively. For a stage where $T_s(d2)$ and $T_s(d3)$ are equal, we call it a *static stage*; other stages are called *dynamic stages*.

For a static stage, therefore, $T_s(d1)$ should be equal to $T_s(d2)$ or $T_s(d3)$, and the $e_s(d1)$ should be the average of task execution times $e_s(d2)$ and $e_s(d3)$. However to reduce the model solution time, we use the following approximation: (i) we approximate the $T_s(d1)$ to be 1, and $e_s(d1)$ to be the average of the stage execution times (obtained from the spark history server) of the two reference cases corresponding to $d2$ and $d3$, *instead* of their task execution times.

Algorithm 1. Model Solution Algorithm

Input: C, W, S, For each $s \in S : Succ(s), Pred(s), T_s, e_s$
Output: T_{exec}
1: $RS, RT, ET, CS \leftarrow \phi$
2: **for each** stage $s \in S$ **do**
3:     **for each** task t of stage s **do**
4:         $t.ret \leftarrow e_s$
5:     **end for**
6: **end for**
7: $RS \leftarrow RS \cup \{s$ such that $Pred(s)$ is empty$\}$
8: **for each** stage $s \in RS$ **do**
9:     **for each** task t of stage s **do**
10:         $RT \leftarrow RT \cup \{t\}$
11:     **end for**
12: **end for**
13: $T_{exec} \leftarrow 0$
14: $T_{next} \leftarrow 0$
15: **repeat**
16:     **for each** available core $c \in C$ **do**
17:         **if** RS is not empty **then**
18:             stage $s \leftarrow$ next stage from RS
19:             task $t \leftarrow$ next task of stage s to be scheduled
20:             $RT \leftarrow RT - \{t\}$
21:             $ET \leftarrow ET \cup \{t\}$
22:             Schedule task t in core c
23:         **end if**
24:     **end for**
25:     $T_{next} \leftarrow min\{t.ret : t \in ET\}$
26:     $T_{exec} \leftarrow T_{exec} + T_{next}$
27:     **for each** task $t \in ET$ **do**
28:         **if** $t.ret > T_{next}$ **then**
29:             $t.ret \leftarrow t.ret - T_{next}$
30:         **else**
31:             task t is completed
32:             $ET \leftarrow ET - \{t\}$
33:         **end if**
34:     **end for**
35:     **for each** stage $s \in RS$ **do**
36:         **if** all the tasks of stage s have completed **then**
37:             $CS \leftarrow CS \cup \{s\}$
38:             $RS \leftarrow RS - \{s\}$
39:         **end if**
40:     **end for**
41:     **for each** stage $s \in CS$ **do**
42:         **for each** $v \in Succ(s)$ of stage s **do**
43:             **if** all $u \in Pred(v)$ have completed **then**
44:                 $RS \leftarrow RS \cup \{v\}$
45:                 **for each** task t of stage v **do**
46:                     $RT \leftarrow RT \cup \{t\}$
47:                 **end for**
48:             **end if**
49:         **end for**
50:     **end for**
51:     $CS \leftarrow \phi$
52: **until** RS is empty
53: $T_{exec} \leftarrow T_{exec} + W$ **return** T_{exec}

For a dynamic stage, we have observed that the number of tasks in a stage always increases with the increase in input size, and this increase is roughly *linear*. Thus, the $T_s(d2)$ and $T_s(d3)$ are dependent on input data sizes and $T_s(d3)$ is greater than $T_s(d2)$ ($d3$ being greater than $d2$). For such a stage, we estimate $e_s(d1)$ as $e_s(d1) = (e_s(d2) + e_s(d3))/2$, and $T_s(d1)$ as

$$T_s(d1) = T_s(d3) + [[(T_s(d3) - T_s(d2))/(d3 - d2)] * (d1 - d3)] \qquad (1)$$

Estimation of W: We first run our model for the input data size $d2$ with warm-up time as 0, and obtain the predicted execution time of the application $T_{exec}(d2)$ from the model. Next, we run the model for the input data size $d3$ with warm-up time as 0, obtain the predicted execution time of the application $T_{exec}(d3)$ from the model. Let the measured execution time of the application for the input data size $d2$ be $M_{exec}(d2)$ and that for the input data size $d3$ be $M_{exec}(d3)$. Note that the $M_{exec}(d2)$ and $M_{exec}(d3)$ are the average of four runs, each for $d2$ and $d3$ respectively. Subsequently, we estimate the warm-up time to be utilized for predicting the application's execution time for an arbitrary input data size such as $d1$ as

$$W(d1) = (|M_{exec}(d2) - T_{exec}(d2)| + |M_{exec}(d3) - T_{exec}(d3)|)/2 \qquad (2)$$

4   Results and Analysis

In this section, first, we illustrate our proposed model solution algorithm through a synthetic DAG shown in Fig. 3; second, we validate our model against the measurement results of a big data query, Query-64, of the TPC-DS benchmark. In each case, we compare our model against the adapted model of Shah et al. [9] where relevant. All the models are Python 3 programs that are run using Jupyter Notebook on a machine with following specification: Intel(R) Core(TM) i7-8550U CPU @ 1.80 GHz, 2.00 GHz Processor and 16 GB RAM.

4.1   Walk-through of Model Solution Algorithm Using Synthetic DAG

In this section we explain our model solution algorithm using the synthetic DAG of Fig. 3. Table 1 presents the details on the stages of the DAG.

The first column of Table 1 shows every stage of the DAG, while the second and third columns show the predecessors and successors of each stage. The ϕ indicates absence of predecessors/successors of a stage. If free executors are available, we see here an opportunity to run stage 5 even before stage 4 has completed. This is because stage 5 has no dependency on stage 4. Former approaches [1,9] would have waited for all the parallel stages 2, 3 and 4 to finish before starting the execution of stage 5. This is where our model overcomes the limitation of the previous works.

Table 1 further shows the number of tasks for a stage, and the average task execution time for that stage, in the two rightmost columns respectively. We consider 8 executor cores (i.e. C is 8) which implies that a maximum of 8 tasks can run in parallel at any given point of time. For this example, we assume the warm-up time W of the application for starting the driver and the executors to be 0. Our model solution algorithm predicts the execution time T_{exec} of the DAG of Fig. 3 to be 11 s.

Table 1. Details of the DAG of Fig. 3—every stage, it's predecessor(s), successor(s), number of tasks of the stage, average task execution time for the stage. The ϕ indicates absence of predecessors/successors of a stage.

Stage s	Predecessors $Pred(s)$	Successors $Succ(s)$	No. of Tasks T_s	Average Task Execution Time e_s (sec)
1	ϕ	2, 3, 4	1	1
2	1	5	2	2
3	1	5	5	2
4	1	6	3	5
5	2, 3	6	2	4
6	4, 5	ϕ	2	2

Table 2 depicts the progress of our model solution algorithm (i.e. Algorithm 1) on the DAG of Fig. 3. In the table, the "Algo Lines" column shows the nth iteration of the algorithm and the range of line numbers present in the description of the algorithm (see Algorithm 1 for the line numbers). The row 1 shows the status of the variables RS, RT, ET, CS, T_{next} and T_{exec}—This row indicates that task 1–1 of stage 1 is ready to be scheduled. Row 2 shows that task 1–1 is scheduled to one of the eight cores and is currently executing. Row 3 shows that the task 1–1 is complete after 1 s (T_{next} column), and consequently stage 1 is complete; the tasks of stages 2, 3, 4 are ready to be scheduled. Next, in row 4, eight of the tasks of stages 2, 3, 4 are scheduled to the eight available cores in a round-robin fashion; since all eight cores get busy, the two tasks 3–4 and 3–5 still remains as ready to be scheduled. The algorithm thus continues for 6 iterations, and eventually predicts the execution time T_{exec} of the DAG of Fig. 3 to be 11 s.

Table 3 shows the impact of varying average task execution time e_s of stage 4 (a parallel stage) on the execution time T_{exec} of the DAG of Fig. 3, while keeping the number of tasks of stage 4 constant. To compare against our model, we adapt the model of Shah et al. [9] and execute it with assumed values of the model input parameters that are same as those we use for our model.

We find that as the e_s of stage 4 is increased from 5 s to 7 s to 9 s, Shah model [9] predicts T_{exec} that is found increasingly greater compared to our model—percent of T_{exec} difference increasing from 9.09 to 27.27 to 33.33. This is because Shah model follows the philosophy of sequential scheduling and hence is unable

Table 2. Progress of model solution algorithm on the DAG of Fig. 3

Algo Lines	RS	RT	ET	CS	T_{next}	T_{exec}
1–14	1	1–1	ϕ	ϕ	0	0
Iter-1: 16–24	1	ϕ	1–1	ϕ	0	0
Iter-1: 25–50	2, 3, 4	2–1, 2–2, 3–1, 3–2, 3–3, 3–4, 3–5, 4–1, 4–2, 4–3	ϕ	1	1	1
Iter-2: 16–24	2, 3, 4	3–4, 3–5	2–1, 3–1, 4–1, 2–2, 3–2, 4–2, 3–3, 4–3	ϕ	1	1
Iter-2: 25–50	3, 4	3–4, 3–5	4–1, 4–2, 4–3	2	2	3
Iter-3: 16–24	3, 4	ϕ	4–1, 4–2, 4–3, 3–4, 3–5	ϕ	2	3
Iter-3: 25–50	4, 5	5–1, 5–2	4–1, 4–2, 4–3	3	2	5
Iter-4: 16–24	4, 5	ϕ	4–1, 4–2, 4–3, 5–1, 5–2	ϕ	2	5
Iter-4: 25–50	5	ϕ	5–1, 5–2	4	1	6
Iter-5: 16–24	5	ϕ	5–1, 5–2	ϕ	1	6
Iter-5: 25–50	6	6–1, 6–2	ϕ	5	3	9
Iter-6: 16–24	6	ϕ	6–1, 6–2	ϕ	3	9
Iter-6: 25–50	ϕ	ϕ	ϕ	6	2	11

to schedule stage 5 before stage 4 completes, whereas our model, being able to account for stage parallelism, is able to schedule stage 5 as soon as stage 2 and 3 are complete, regardless of stage 4 completion.

The two rightmost columns of Table 3 shows the time taken to run the two models. We observe that, to run the aforementioned DAG, our model takes time that is similar to that of Shah model.

4.2   Model Validation Using a Big Data Query—Query-64

In this section, we validate our model against the measured execution time obtained by running a big data query, Query-64, on the spark cluster of the Google Cloud. Query-64 is a complex query in TPC-DS benchmark (a popular big data benchmark [10]) that involves parallel execution of a large number of stages.

Our goal here is to predict the execution time of Query-64 for an input data size of 200 GB on eight executor cores. To estimate the model parameter values for this input data size (i.e. 200 GB), we run the query for two data sizes 20 GB and 100 GB as our two reference cases. With respect to each case, we run the spark application (i.e. Query-64) four times and then, for each stage of the

DAG of Query-64, we take the average of the task execution times obtained from those four runs corresponding to the case. We carry out this experiment in Google Cloud, on a spark cluster consisting of one master and two worker nodes, each node being an n1-standard-4 machine with primary disk size of 500 GB, 4 virtual CPU cores, and a Linux image (version 2.0-debian10) containing the spark framework.

Table 3. Impact on T_{exec} on varying average task execution time e_s of stage 4

Stage-4 e_s (s)	Shah model [9] T_{exec} (s)	Our model T_{exec} (s)	T_{exec} diff (%)	Shah model [9] solution time (ms)	Our Model solution time (ms)
5	12	11	9.09	1	2
7	14	11	27.27	1	1
9	16	12	33.33	1	1

Table 4 shows the predecessors and successors of each stage of the DAG of the application Query-64. Similar to Table 1, the first three columns of the table show each stage of the DAG, its predecessor stages, and its successor stages respectively. We use the information related to the reference cases of 20 GB and 100 GB to estimate the values of the input parameters for 200 GB input data. We follow the steps provided in Sect. 3.3 to accomplish this estimation. The two rightmost columns of Table 4 shows the *estimated* number of tasks T_s and the *estimated* average task execution time e_s for each stage. The warm-up time W for Query-64 is estimated as 32.11 s.

The measured execution time of the spark application Query-64 for 200 GB input data is approximately 491.71 s (which is an average of four measurements, the standard deviation being approximately 11.08 s). Next, we develop a model for the DAG of Query-64. On running our model solution algorithm using the parameter values of Table 4, we obtain the predicted execution time T_{exec} to be 477.71 s, that is, an absolute error value of $(|(491.71 - 477.71)|/491.71) \times 100\% = 2.85\%$ with respect to the measured data. Further, we adapt the model of Shah et al. [9] and execute it with the estimated values of the model input parameters which are same as that of our model. Comparing its results against measured data reveals an error of 7.21% that is *larger* than our model's error.

5   Related Work

The works in [5,7,11] modelled spark applications using machine learning techniques while those in [2] and [3] modelled such applications using queuing networks and simulation modelling techniques. Unlike our work, an assumption in these works was that all the spark jobs are always sequentially scheduled although some of them may have the potential to execute in parallel in reality. Besides, the aforementioned modeling techniques require domain expertise and

Table 4. Details of the DAG of Query-64—every stage, it's predecessor(s), successor(s), number of tasks of the stage, average task execution time for the stage. The ϕ indicates absence of predecessors/successors of a stage. The input data size is 200 GB.

Stage s	Predecessors $Pred(s)$	Successors $Succ(s)$	Estimated No. of Tasks, T_s	Estimated average task execution time e_s (sec)
0	ϕ	1, 2, 3, 4, 5, 6, 7, 8, 9 ,10, 11	1	12.375
1	0	12	1	2.75
2	0	12	8	0.40625
3	0	12	1	2.25
4	0	12	1	1.85
5	0	12	1	1.5625
6	0	12	1	2.1
7	0	12	1	1.17625
8	0	12	1	1.15
9	0	12	3	1.475
10	0	12	9	1.65
11	0	12	8	1.0625
12	1, 2, 3, 4, 5, 6, 7, 8, 9, 10, 11	13	1	7.625
13	12	14	1	6
14	13	15, 16, 17, 18, 22	1	6.625
15	14	20	90	5.875
16	14	20	52	6.125
17	14	19	83	5.25
18	14	19	224	5.375
19	17, 18	21	1	28.5
20	15, 16	21	1	29.25
21	19, 20	25	1	12.75
22	14	23	52	6.125
23	22	24	1	35.875
24	23	25	1	13.25
25	21, 24	26	1	1.375
26	25	27	1	1.375
27	26	28	1	0.7375
28	27	ϕ	1	1

incur substantial model development time and financial cost. Our technique is a simpler one that is more convenient to model a system exhibiting low variability in its execution.

The work of Shah et al. [9] proposed a deterministic model to predict application execution time. It proposed a concept called *groups* to categorize the parallel stages of a job, yet it *enforced* a condition that all the stages in a group *must* finish before a succeeding dependent stage can start although the succeeding stage may not be dependent on all the stages of the preceding group. As a result, the succeeding stage will not be able to start even if the stages it depends upon have already finished and executor cores are available for execution.

6   Discussions

Our work describes a methodology with an algorithm for the deterministic prediction of Spark-based applications' execution times. Given that one, the execution times of the tasks of a big-data application's DAG have an almost deterministic duration in many practical cases, and two, the round-robin scheduling is used in many big-data frameworks, our model is able to predict accurate results assuming these two factors. Yet, our work is limited in multiple aspects—one, we assume full concurrency between the considered executors. Also, the presence of high variability, or advanced scheduling techniques in computing systems can diminish the ability of our algorithm to perform accurate predictions.

Nevertheless, the result of this work can act as a guidance; in practice, there could be systems where duration of task execution times and scheduling choices are almost deterministic. In such cases, it might be worthwhile applying simple deterministic techniques—like the one proposed here, having an acceptable trade-off between execution time and accuracy—instead of applying complex simulation or analytical computations.

7   Conclusion and Future Work

In this paper we have developed a graph-based, deterministic analytical model that predicts the execution time of spark applications. Unlike earlier models which follow the philosophy of sequential scheduling of jobs, our model is able to account for the parallelism in spark job executions. We have validated our model against the measured execution time obtained by running a big data query (Query-64 of TPC-DS benchmark) that involves parallel execution of a large number of stages. The query was executed on the spark cluster of Google Cloud. Our model resulted in an execution time that is at 2.85% error in comparison to the measured execution time. Our work has considered fixed number of executors allocated for the spark application. In future we plan to model a scenario that involves dynamic allocation of executors, that is, where the executors are allocated on demand for the application at run-time. Besides, we plan to investigate how our model can be combined with an optimization framework to aid in the decision making for optimal number of executors to be allocated for the application.

References

1. Amannejad, Y., Shah, S., Krishnamurthy, D., Wang, M.: Fast and lightweight execution time predictions for spark applications. In: 2019 IEEE 12th International Conference on Cloud Computing (CLOUD), pp. 493–495 (2019)

2. Ardagna, D., et al.: Performance prediction of cloud-based big data applications. In: 2018 ACM/SPEC 9th International Conference on Performance Engineering (ICPE), pp. 192–199 (2018)
3. Ardagna, D., et al.: Predicting the performance of big data applications on the cloud. J. Supercomput. **77**, 1321–1353 (2021)
4. Asaadi, H., Khaldi, D., Chapman, B.: A comparative survey of the HPC and big data paradigms: Analysis and experiments. In: 2016 IEEE International Conference on Cluster Computing (CLUSTER), pp. 423–432 (2016)
5. Didona, D., Quaglia, F., Romano, P., Torre, E.: Enhancing performance prediction robustness by combining analytical modeling and machine learning. In: 2015 ACM/SPEC 6th International Conference on Performance Engineering (ICPE), pp. 145–156 (2015)
6. Karau, H., Konwinski, A., Wendell, P., Zaharia, M.: Learning spark: lightning-fast big data analysis, O'Reilly Media Inc (2015)
7. Maros, A., et al.: Machine learning for performance prediction of spark cloud applications. In: 2019 IEEE 12th International Conference on Cloud Computing (CLOUD), pp. 99–106 (2019)
8. Nguyen, N., Khan, M., Albayram, Y., Wang, K.: Understanding the influence of configuration settings: an execution model-driven framework for Apache spark platform. In: 2017 IEEE 10th International Conference on Cloud Computing (CLOUD), pp. 802–807 (2017)
9. Shah, S., Amannejad, Y., Krishnamurthy, D., Wang, M.: Quick execution time predictions for spark applications. In: 2019 IEEE 15th International Conference on Network and Service Management (CNSM), pp. 1–9 (2019)
10. TPC-DS decision support benchmark. www.tpc.org/tpcds/
11. Venkataraman, S., Yang, Z., Franklin, M., Recht, B., Stoica, I. : Ernest: efficient performance prediction for large-scale advanced analytics. In: 13th USENIX Symposium on Networked Systems Design and Implementation NSDI 2016, pp. 363–378 (2016)
12. Wang, K., Khan, M.: Performance prediction for apache spark platform. In: 2015 IEEE 17th International Conference on High Performance Computing and Communications, 2015 IEEE 7th International Symposium on Cyberspace Safety and Security, and 2015 IEEE 12th International Conference on Embedded Software and Systems, pp. 166–173 (2015)

Modelling Paradigms and Tools

A Scalable Opinion Dynamics Model Based on the Markovian Agent Paradigm

Marco Scarpa$^{(\boxtimes)}$ (iD), Salvatore Serrano (iD), and Francesco Longo (iD)

Università degli Studi di Messina, Messina, Italy
{mscarpa,sserrano,flongo}@unime.it

Abstract. Agent-based approaches are frequently used in the literature for the analytical modeling of opinion dynamics. In the cases in which each agent is modeled through a Markov chain, state-space explosion arises as main limit to the scalability of the model. Lumping or simulation are often the only alternatives. In this paper, we propose the use of Markovian Agents to model Peer Assembly scenarios in which agents influence each other in an atomic and linear way. Results show how this modeling paradigm is effective and flexible in providing transient and steady-state solutions for several conditions of parameter settings. This opens the way for future work in modeling more complex scenarios such as gossip-based interactions or the presence of stubborn agents.

Keywords: Markovian Agents · Opinion dynamics · Stochastic modelling · Markov chains

1 Introduction

In a fully connected world, virtual interactions between people, services, and automatic systems have grown exponentially and affect every-day lives in several contexts: social, political, economical behaviors are strongly influenced by how trust, reputation, and opinions are experienced and shared online [11]. An established trend in literature, even preceding the advent of the Internet era, has been to observe the phenomenon of social interactions from the point of view of the so called social networks, defined as group of individuals bound together by some short or long-term relationships, especially focusing on how opinion dynamics is influenced by the structure and characteristics of these real or virtual networks [8]. Of course, such a way of approaching the problem has applications not only in social and political contexts, but also in finance and business [12], medicine and health care [10], and several others [5].

An interesting research area in this context is related to the use of agent-based approaches for the analytical modeling of opinion dynamics in social networks. In this kind of methods, each individual in the network is modeled as a separated agent while the interactions are represented through a graph [7]. Of particular interest is the way the opinion of each agent is represented in the model. If the opinion can be represented as a numerical value in a discrete domain, Markov chains can be exploited in an effective way [1,3].

In approaches based on the use of Markov chain-based models, the main issue is the state-space explosion. In the literature, a frequent solution is resorting to lumping of the state space with the aim of finding treatable analytic closed forms [3]. Lumping approaches are applicable only when a certain degree of model symmetry is present, e.g., identical agents, identical opinion dynamics for each agent, symmetrical network topology, such as in the Peer Assembly model where the graph representing agent interactions is a full mesh. This limits the variety of scenarios that can be treated. Of course, resorting to simulation-based solution of the model is a frequent approach to solve such an issue, with known drawbacks in terms of run-time costs and solution uncertainty.

The aim of this work is to experiment the Markovian Agent paradigm [2,6] for modelling social interactions among a high number of individuals. We designed our model under the following general hypotheses: (i) *atomic interaction model*: the opinion of each agent is influenced by the opinion of his/her neighbors and this translates in the model in the fact that the state transition rates in the Markov chain model of each agent depend on the state of the neighbors; (ii) *linear emulative model*: specifically, the transition rate from one opinion to a different one is linearly proportional to the number of neighbors that share that opinion.

Even if the model we discuss here is about a quite simple kind of interaction, we will show its effectiveness and flexibility by providing transient and steady-state solutions in several scenarios. First, we will show results considering different influence rates and fixed (small) number of agents. Then, we will fix the influence rate and consider growing number of agents. Finally, scenarios with several influence rates and fixed (large) number of agents will be analyzed. This is the starting point for developing models for more complex scenarios, probably difficult to be studied except with simulation. We implement and solve the proposed model by exploiting MAGNET [9], a C library implementing all the primitives for creating multi-class Markovian Agent analytical models with message-based interactions, and numerically solving the set of differential equations describing the dynamics of the system.

The paper is organized as follows. In Sect. 2, we present an overview of Markovian Agents modelling paradigm, while in Sect. 3, we propose the interaction model for the considered opinion dynamics scenario. Results obtained through analysis of the proposed model are presented in Sect. 4, while Sect. 5 concludes the paper with some hints on possible future work.

2   The Markovian Agent Modelling Paradigm

Markovian Agent Models (MAMs) represent systems as a collection of agents scattered over a geographical space, and described by continuous-time Markov chains (CTMC) where two types of transitions may occur: *local transitions* and *induced transitions*. The former models the internal features of the Markovian

Agent (MA), whereas the latter accounts for interaction with other MAs based on their states. Whereas the original formulation of the Markovian Agent modelling paradigm [6] implements agents interaction with message exchange, it has been generalized in [2] in order to encompass different forms of dependencies among the involved entities.

In this paper, we exploit the message passing interaction paradigm because it is meaningful in representing the interaction we deal with, thus we quickly summarize the MAM analytical formulation focusing only on it. An MA can send a message to other MAs either when a local transition occurs or remaining into a specific state. The propagation of messages is regulated by the *perception function* $u(\cdot)$. Depending on the agent position in the space, on the message routing policy, and on the transmittance properties of the medium, this function allows the receiving MA to be aware of the state from which the message was issued, and to use this information to choose an appropriate action. MAs can be scattered over a geographical area $\mathcal{V}$. Agents can be grouped in classes and can share different types of messages.

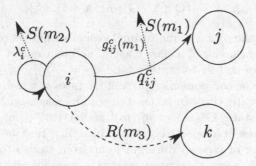

Fig. 1. Graphical representation of MAs

We represent a Markovian agent by exploiting the graphical notation shown in Fig. 1. Given an MA of class c, a local transition from state i to state j is drawn as usual with a solid arc and the associated rate q_{ij}^c. When a transition happens a message m could be sent with probability $g_{ij}^c(m)$; this event is graphically drawn as a dotted line starting from the transition whose firing sends the message and it is labeled with $S(m)$ to make evident which message is sent. An MA is also able to send a message during the sojourn in a particular state: self-loops are used to this aim; in fact, a self-loop in state i could be used to send a message m at a given rate λ_i^c similarly to what happens during a state transition. It is worth noting that self-loops do not influence local behavior of MAs, like in the usual theory of CTMCs, due to the memoryless property of the exponential distribution; instead they have a role in the evolution of a remote MA receiving the corresponding message. An example of MA is depicted in Fig. 1, where message m_1 is sent at the occurrence of transition from i to j, and a self-loop is associated with state i

emitting message m_2 at rate λ_i^c. Induced transition due to reception of a message is graphically represented with a dashed arc between involved states; in this case the arc is labeled with $R(m)$. As an example, in Fig. 1, a transition from state i to state k, is due to the reception of message m_3.

Formally a *Multiple Agent Class, Multiple Message Type MAM* is defined by the tuple:

$$MAM = \{\mathcal{C}, \mathcal{M}, \mathcal{V}, \mathcal{U}, \mathcal{R}\}, \tag{1}$$

where $\mathcal{C} = \{1 \ldots C\}$ is the set of agent classes, and $\mathcal{M} = \{1 \ldots M\}$ is the set of message types. $\mathcal{V}$ is the finite space over which MAs are spread, and $\mathcal{U} = \{u_1(\cdot) \ldots u_M(\cdot)\}$ is a set of M perception functions (one for each message type). The density of the agents is regulated by functions $\mathcal{R} = \{\xi^1(\cdot) \ldots \xi^C(\cdot)\}$, where each component $\xi^c(\mathbf{v})$, with $c \in \mathcal{C}$, counts the number of class c agents deployed in position $\mathbf{v} \in \mathcal{V}$. Since in this work the space is considered discrete, each position could be identified by a *cell* numbered with an integer with respect to some reference system.

Each agent MA^c of class c is defined by the tuple:

$$MA^c = \{\mathbf{Q}^c, \mathbf{\Lambda}^c, \mathbf{G}^c(m), \mathbf{A}^c(m), \boldsymbol{\pi}_0^c\}. \tag{2}$$

Here, $\mathbf{Q}^c = [q_{ij}^c]$ is the $n_c \times n_c$ infinitesimal generator matrix of the CTMC that describes the local behavior of a class c agent, and its element q_{ij}^c represents the transition rate from state i to state j (and $q_{ii}^c = -\sum_{j \neq i} q_{ij}^c$). $\mathbf{\Lambda}^c = [\lambda_i^c]$, is a vector of size n_c whose components represent the rates at which the Markov chain reenters the same state: this can be used to send messages with an assigned rate without leaving a state. $\mathbf{G}^c(m) = [g_{ij}^c(m)]$ and $\mathbf{A}^c(m) = [a_{ij}^c(m)]$ are $n_c \times n_c$ matrices that represent respectively the probability that an agent of class c generates a message of type m during a jump from state i to state j, and the probability that an agent of class c accepts a message of type m in state i and immediately jumps to state j. $\boldsymbol{\pi}_0^c$, is a probability vector of size n_c which represents the initial state distribution.

The perception function of a MAM is formally defined as $u_m : \mathcal{V} \times \mathcal{C} \times \mathbb{N} \times \mathcal{V} \times \mathcal{C} \times \mathbb{N} \to \mathbb{R}^+$. The values of $u_m(\mathbf{v}, c, i, \mathbf{v}', c', i')$ represent the probability that an agent of class c, in position $\mathbf{v}$, and in state i, perceives a message m generated by an agent of class c' in position $\mathbf{v}'$ in state i'. Thanks to perception functions, different instances of agents deployed over the space can interact sending messages one each other. Interactions are technically implemented through the matrix $\mathbf{\Gamma}^c(t, \mathbf{v}, m)$, a diagonal matrix collecting the total rate of received messages m by an agent of class c in position $\mathbf{v}$ (element γ_{ii} stores the value for state i).

$\mathbf{\Gamma}^c(t, \mathbf{v}, m)$ elements are computed as in the following. $\beta_j^c(m)$ stores the total rate at which messages of type m are generated by an agent of class c in state j, and it is evaluated as

$$\beta_j^c(m) = \lambda_j^c \, g_{jj}^c(m) + \sum_{k \neq j} q_{jk}^c \, g_{jk}^c(m). \tag{3}$$

Let us denote with $\boldsymbol{\pi}_j^c(t, \mathbf{v})$ the probability vector of the class c agent in position $\mathbf{v}$; the total rate of messages received by an agent of class c, in state i, at position $\mathbf{v}$, at time t is

$$\gamma_{ii}^c(t, \mathbf{v}, m) = \sum_{\mathbf{v}'} \sum_{c'=1}^{C} \sum_{j=1}^{n_{c'}} u_m(\mathbf{v}, c, i, \mathbf{v}', c', j)\beta_j^{c'}(m)\boldsymbol{\pi}_j^{c'}(t, \mathbf{v}). \tag{4}$$

The rates in (4) are collected in a diagonal matrix $\boldsymbol{\Gamma}^c(t, \mathbf{v}, m) = \mathrm{diag}(\gamma_{ii}^c(t, \mathbf{v}, m))$ that is used to compute the kernel (infinitesimal generator matrix) of class c agent at position $\mathbf{v}$ at time t:

$$\mathbf{K}^c(t, \mathbf{v}) = \mathbf{Q}^c + \sum_m \boldsymbol{\Gamma}^c(t, \mathbf{v}, m) \left[\mathbf{A}^c(m) - \mathbf{I}\right]. \tag{5}$$

The overall MA model thus evolves according to the set of coupled differential equations

$$\frac{d\boldsymbol{\pi}^c(t, \mathbf{v})}{dt} = \boldsymbol{\pi}^c(t, \mathbf{v})\mathbf{K}^c(t, \mathbf{v}) \tag{6}$$

under the initial condition $\boldsymbol{\pi}_0^c$, $\forall \mathbf{v} \in \mathcal{V}$, $\forall c \in \mathcal{C}$.

As deeply described in [4,6], the main advantage of MA is that state space complexity is maintained low because dependencies between two agents are modeled through messages instead of defining the cross product of their state spaces.

3 Model of Interacting Individuals

3.1 Reference Interaction Model

In this section we summarizes the reference interaction model we deal with.

A social network with its interactions is usually represented as an undirected graph, where the nodes are the individuals and the edges connecting two nodes represent the interaction between two individuals; moreover, each node is also characterized by a position in the "virtual" space[1]. The influence is considered reciprocal. We assume the social network is made by N individuals, and each of them could have an opinion in the set $\mathcal{O}$, whose multiplicity is $|\mathcal{O}|$. Each individual has a standalone behavior, according to which the opinion can change according to a continuous time Markov chain. We denote the state of an individual in position $\mathbf{v}$ at time t with $\sigma(t, \mathbf{v})$, the standalone infinitesimal generator matrix with $\mathbf{Q}(\mathbf{v})$, and define the state probability at time t as $\pi_i(t, \mathbf{v}) = P\{\sigma(t, \mathbf{v}) = i\}$. Similarly, $\sigma(\mathbf{v})$ and $\pi_i(\mathbf{v}) = P\{\sigma(\mathbf{v}) = i\}$ denote the state of an individual in position $\mathbf{v}$ and its corresponding probability in stable condition respectively.

Let us consider an individual in position $\mathbf{v}$; its behavior is influenced by its neighbors that push it to change opinion based on their own current opinion. In this work we assume an individual influences its neighbors boosting the opinion

---

[1] We use the term "virtual" because the space could be either the physical space or not, like in applications where the geographical position is not relevant.

change; as a consequence, the matrix $\mathbf{Q}(\mathbf{v})$ of the individual in position $\mathbf{v}$ changes based on the states of the neighbors, and it has to be replaced by $\mathbf{Q}'(\mathbf{v}, t)$ whose values depend on the neighbors' state at time t. Such a kind of interaction is known as *atomic interaction*. In the atomic interaction model, the standalone behavior and the neighbors influences sum up, thus we have $\mathbf{Q}'(\mathbf{v}, t) = \mathbf{Q}(\mathbf{v}) + \mathbf{I}(\mathbf{v}, t)$, where $\mathbf{I}(\mathbf{v}, t)$ depends on the kind of interaction considered in the social network.

Let $\eta(\mathbf{v})$ be the number of neighbors of the individual in position $\mathbf{v}$, and $\Upsilon(\mathbf{v})$ the set of positions of individuals neighboring with individual in position $\mathbf{v}$; in this work we will focus on interactions such that the matrix $\mathbf{I}(\mathbf{v}, t)$ is defined in the following Eq. (7)

$$I_{i,j} = \begin{cases} \eta(\mathbf{v})\lambda_j \sum_{v' \in \Upsilon(\mathbf{v})} \delta_j(t, \mathbf{v}'), & i \neq j \\ -\sum_{j:j \neq i} \Gamma_{i,j}(t, \mathbf{v}), & i = j \end{cases}, \quad \delta_j(t, \mathbf{v}) = \begin{cases} 1, & \sigma(t, \mathbf{v}) = j \\ 0, & \text{otherwise} \end{cases}. \quad (7)$$

This latter case of interaction is known in literature as *linear emulative*. The linear emulative model assumes that the interaction rate to change to state j an individual is subject to is proportional to the number of neighboring individuals in the state j.

The opinion dynamic over the time depends on both the network topology and the individual behaviors as well as the influence strength of each single individual in the network. In this work we will focus on the *Peer Assembly* where all the individuals are equal and their initial opinion is an independent and identically distributed random variable; moreover the network is made by N individuals fully interconnected, i.e. each individual communicate with all the others. Peer Assembly is often used as reference scenario due to symmetries it implies, especially when stochastic based modelling is used; even if the resulting model could be simple in its structure, it is good for representing some real cases. Since our purpose is to experiment the Markovian Agent paradigm, we also use the Peer Assembly scenario in this work, but the use of Markovian Agent opens to easily create more complex scenarios where the involved agents possibly have different structures differently by the other stochastic approaches discussed in literature.

3.2   The Markovian Agent Model of Peer Assembly

In the modeling approach we are proposing, we suppose, like in [3], each individual is characterized by two discrete opinions and each individual may change its opinion spontaneously. We thus model the set of interacting individuals with one class only, because individual behaviors are considered identical. We denoted such a class as *Individual* and we use a superscript I for identifying all its quantities as defined in Eq. (2). An agent in the class is made by two states representing the two opinions each individual can have at a given time. The spontaneous change of opinion when an individual is isolated is represented by rates between the two states, q_{ij}, with $i \neq j$ and $i, j = 1, 0$, under the assumption individuals opinion evolves according to a continuous-time Markov chain. The matrix

$\mathbf{Q}^I$ will denote the infinitesimal generation matrix of the class. It is worth to note that $\mathbf{Q}^I$ is the $\mathbf{Q}(\mathbf{v})$ introduced in Sect. 3.1. Both the two states are also characterized with self-loops, whose rate in state i is λ_i, $i = 1, 0$. Thanks to the self loops, an agent in state i emits a message m_i. We use messages emitted by self-loops to represent the influence an individual carries out on other individuals with an influence strength intensity λ_i. The class *Individual* is depicted in Fig. 2, where the graphical notation introduced in Sect. 2 has been used.

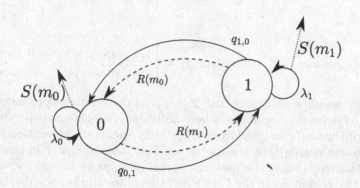

Fig. 2. The Markovian Agent class modelling opinion dynamics of a standalone entity

According to the description given above, matrices $\mathbf{Q}^I$ and $\mathbf{\Gamma}^I$ of the class *Individual* are:

$$\mathbf{Q}^I = \begin{bmatrix} -q_{01} & q_{01} \\ q_{10} & -q_{10} \end{bmatrix}, \quad \mathbf{\Gamma}^I = \begin{bmatrix} \lambda_0 & 0 \\ 0 & \lambda_1 \end{bmatrix} \tag{8}$$

We lastly model the opinion change due to the influence of other neighbors' by exploiting induced transitions. Induced transitions between the two states triggered by the perceiving of messages m_0 and m_1 are set. In such a way when an individual is in the state 0 (1) and perceives a message m_1 (m_0), because a neighboring individual has a different opinion and the corresponding agent sends the message characterizing it, the natural transition rate q_{01} (q_{10}) to the other state is strengthened with the influence of the neighbor. In this first proposal we assume that the influence due to neighbors is not filtered in any way, thus we set the acceptance probability in the class to 1.0 for both the induced transitions. The acceptance matrices introduced in Sect. 2 are:

$$\mathbf{A}^I(m_0) = \begin{bmatrix} 0.0 & 0.0 \\ 1.0 & 0.0 \end{bmatrix}, \quad \mathbf{A}^I(m_1) = \begin{bmatrix} 0.0 & 1.0 \\ 0.0 & 0.0 \end{bmatrix} \tag{9}$$

The Markovian agent class *Individual* describes the behavior of each agent in standalone where the induced transitions are possible, but they actually depends on the state of the neighbors. The real influence has to be defined describing who are the neighbors of each individual. The perception functions are exploited to this purpose. Since we want to study a Peer Assembly, each individual is considered a neighbor of all the others and the perception functions for m_0 and m_1 set a probability $p = 1.0$ to perceive a message m_0 when an individual is in the state 1 and a message m_1 when it is in the state 0, irrespective of the spatial position of the couple of individuals:

$$u_{m0}(\mathbf{v}, Individual, i, \mathbf{v}', Individual, j) = \begin{cases} 1.0, & \forall \mathbf{v}, \mathbf{v} \in \mathcal{V}, i = 1, j = 0 \\ 0.0, & \text{otherwise} \end{cases}, \quad (10)$$

and

$$u_{m1}(\mathbf{v}, Individual, i, \mathbf{v}', Individual, j) = \begin{cases} 1.0, & \forall \mathbf{v}, \mathbf{v} \in \mathcal{V}, i = 0, j = 1 \\ 0.0, & \text{otherwise} \end{cases}. \quad (11)$$

This latter assumption means that the spatial distribution does not play any role, but more complex scenarios where individual interactions depends on their position can be easily designed by appropriately setting the perception functions.

It is worth noting that the model we described implicitly implements the linear *emulative behavior* because the kernel of agent in position $\mathbf{v}$, computed with Eq. (5), is built using the $\mathbf{\Gamma}^I(t, \mathbf{v}, m)$ that sums all the contributions of the deployed agents influencing it (see Eq. (4)), similarly to the Eq. (7).

4   Peer Assembly Analysis

In this section, we present and discuss results obtained analyzing the Peer Assembly scenario in several conditions of parameters settings with the aim to show the potentiality of the proposed model in exploring the related opinion dynamics behavior. The model has been implemented in C, with the support of MAGNET [9] C library, that is a tool for the analysis of models based on the Markovian Agent modelling paradigm[2].

4.1   Different Influence Rates and Fixed (Small) Number of MAs

In order to demonstrate basic properties of the proposed model, we carried out some experiments fixing to $N = 3$ the number of MAs. In this simple condition, we investigated the behavior of the model varying the bias of the influence rates. All the individuals in the experiment share the same parameters, i.e., each individual has the same behavior. Specifically, we set the spontaneous change rate $q_{0,1} = q_{1,0} = 0.25$ in each experiment. All the individuals start from the same opinion, accordingly, we set to 1.0 the probability of state 1 and 0 the probability of state 0 at time $t = 0$, i.e. $\boldsymbol{\pi}_0^I = [0.0, 1.0]$. Therefore, we conducted several experiments changing the emissions rate of influence message λ_0^* from 0 to 0.250 with a step of 0.025 and settings $\lambda_1^* = 0.250 - \lambda_0^*$. Results show the trend of the probability of each individual to stay in state 1. Moreover, in order to weight the

---

[2] MAGNET is available at http://perf.unime.it/magnet/.

influence capability with respect to the number of influenced MAs, we assumed three different conditions. In the first one, denoted as configuration $C1$, whose results are depicted in Fig. 3(a), the influence rate is not affected by the number of influenced agents thus we set $\lambda_1 = \lambda_1^*$ and $\lambda_0 = \lambda_0^*$. This condition well reflects a social network behavior, where the opinion posted by an individual is (potentially) spread to all his/her followers regardless of their number. In the second and third one, we assumed the influence rate to be scaled depending on the number of influenced individuals. Specifically, Fig. 3(b) shows results when the influence rate is normalized by the number of possible influenced individuals; this condition, where $\lambda_1 = \frac{\lambda_1^*}{N-1}$ and $\lambda_0 = \frac{\lambda_0^*}{N-1}$, will be denoted as configuration $C2$. Figure 3(c) shows a more aggressive scenario where the influence rate is normalized by the square root of possible influenced individuals and denoted as configuration $C3$, where $\lambda_1 = \frac{\lambda_1^*}{\sqrt{N-1}}$ and $\lambda_0 = \frac{\lambda_0^*}{\sqrt{N-1}}$. The first case well reflects a *one-to-one in presence* influence scenario, where the time spent by each individual in influencing others has to be split depending on the number of other individuals present in the assembly. The second case tries to capture the behavior of a *one-to-many in presence* influence scenario, where each individual is only capable to influence a group of individuals (i.e., by a debate with the travelers in a car or in a public transport vehicle, a debate with a party attendees, and so on) at the same time. Analyzing Fig. 3, it is possible to verify the behavior of individuals is as expected, and the model correctly reflects it. In fact, unbiased influence rates (inducing rate λ_1 greater than λ_0) create an unbalanced condition in favor of the most strengthened opinion, leading to a steady state with higher probability to be in state 1. The condition $\lambda_1 = \lambda_0$ correctly leads to the condition $\pi_0(\mathbf{v}) = \pi_1(\mathbf{v}) = 0.5$, $\forall \mathbf{v} \in \mathcal{V}$, due to the equal spontaneous change rate settings ($q_{0,1} = q_{1,0}$). Comparing results in different conditions, it is possible to highlight how the steady state probabilities are affected by the normalization factor related to the number of affected individuals: in case of Fig. 3(a), where the induction rate is not normalized, the contrast between λ_1 and λ_0 strongly affects the steady state probabilities, while the affection is attenuated in case of Fig. 3(b). The condition in Fig. 3(c), as expected, shows intermediate steady state probabilities compared with the previous ones.

4.2 Fixed Biased Influence Rate and Growing Number of MAs

Having verified the correct behavior of the proposed model for a small number of agents, we carried out some experiments when the number of agents increases. Specifically, we maintains $q_{0,1}$ and $q_{1,0}$ as in the previous experiments, we choose a light biased induction rate couple ($\lambda_1^* = 0.10$ and $\lambda_0^* = 0.15$) and we perform experiments changing N from 3 to 273 at step of 30. We added a new experimental configuration to the three configurations considered in the previous subsection, in which we use a normalization factor based on a threshold Th, with $Th \in \{1, 2, \cdots, N-1\}$, with the aim to establish a superior limit to the interaction rate: if N is lower than Th, we set $\lambda_0 = \lambda_0^*$ and $\lambda_1 = \lambda_1^*$, otherwise we set $\lambda_0 = \lambda_0^* \frac{Th}{N-1}$ and $\lambda_1 = \lambda_1^* \frac{Th}{N-1}$; this latter is denoted as configuration $C4$.

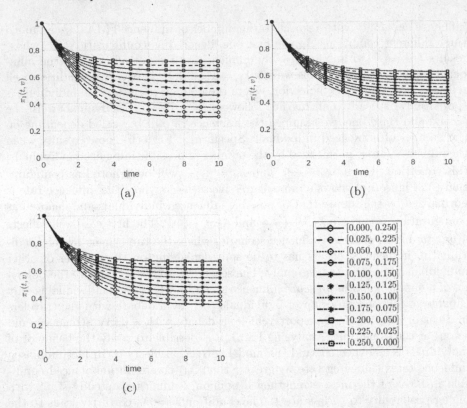

Fig. 3. $\pi_1(t, \mathbf{v})$ with $N = 3$, each couple in the legend $[\lambda_1^*, \lambda_0^*]$ corresponds to different settings of $[\lambda_1, \lambda_0]$ in: (a) configuration *C1*, (b) configuration *C2*, (c) configuration *C3*

Figure 4(a) shows results without any normalization. The steady-state probabilities are strongly affected by the light biased condition when the number of agents grows. This result correctly reflects, in our opinion, the social networks behaviors in which persuasive posted messages suddenly reach all other individuals and a persuasive mechanism based on common opinion is triggered and rapidly grows. Results in Fig. 4(b) shows that, when the inducing rates are normalized with respect to the number of individuals, both steady state and transient are not affected by their number. This behavior, known in literature and highlighted by other models (i.e., [3]), shows that this condition is able to detect only a *one-to-one in presence* influence scenario. Figure 4(c) shows the results for configuration *C3*. As expected, steady-state probabilities are affected by the number of agents to a lesser extent than in the case where there isn't any normalization. In such a condition, an individual is capable to interact only with a portion of the entire population but is able to capture the widespread trend of common opinion. Figure 4(d) and Fig. 4(e) show the behavior of the model with configuration *C4*. The first plot shows the results with $Th = 10$, therefore we have only two different trends for $N = 3$ and $N \geq 33$ respectively. Results in

Fig. 4(e) were obtained using a threshold value $Th = 100$. In this condition, as expected, we have five different trends, the first four obtained when the number of agents is lower then the threshold, i.e. $N = 3, 33, 63, 93$, and the last one when the number of agents is higher then the threshold $N \geq 103$. This condition try to capture the behavior of individuals who are capable to interact with a maximum number of other individuals.

4.3 Several Influence Rates and Fixed (Large) Number of MAs

In order to validate the properties of the proposed model, we carried out the last set of experiments by setting the number of MAs in the model to $N = 273$ and varying the influence rate and the normalization condition like in Subsect. 4.1. We analyzed both the transient of the state probability and the steady-state distribution of number of individuals n_1 in state 1.

Figure 5(a) shows the results when normalization factor is not applied. Three different cases are highlighted: when $\lambda_1 > \lambda_0$ the probability of state 1 approach to 1.0 for all time t; when $\lambda_1 < \lambda_0$ the probability of state 1 suddenly goes to 0.0; if $\lambda_1 = \lambda_0$ the steady state probability becomes 0.5 and the individuals reach the stable condition with a soft negative exponential trend. The steady state distributions of n_1 (Fig. 5(b)) lean towards the value 1 when $\lambda_1 > \lambda_0$ and, at the opposite, they lean over the value 0 when $\lambda_1 < \lambda_0$, both with a very small variance. When $\lambda_1 = \lambda_0$, the mean value of n_1 at steady state becomes $\frac{N}{2}$ with a larger variance.

Applying the normalization factor equal to the number of possible interactions (Fig. 6(a–b)), the state probability moves to steady state condition with an exponential soft trend. The steady state probability will be related to the specific combination of λ_1, λ_0 and, accordingly also the mean value of steady state distribution. The variance of this latter does not appear to be particularly affected by the specific combination of values.

Figure 7(a–b) show the results with the normalization factor equal to the square root of the number of possible interactions. In this condition, the trend of the state probability is affected by the difference of λ_1 and λ_2. Larger is this difference, more suddenly the probability state reaches the steady state. Moreover, as already depicted, the steady state is a function of this values: when $\lambda_1 > \lambda_0$, it approaches 1; at the opposite, it approaches 0, when $\lambda_1 < \lambda_0$. Accordingly, the steady state probability of n_1 will be a mean value approaching N when $\lambda_1 > \lambda_0$ and 0 when $\lambda_1 < \lambda_0$ with a variance inversely proportional to their difference.

Finally, Fig. 8(a–b) show comparisons between different normalization conditions with the same set of parameters used in Subsect. 4.2 and $N = 273$. It is evident that both the conditions with and without the normalization equal to the number of possible interactions correspond to the two limit cases.

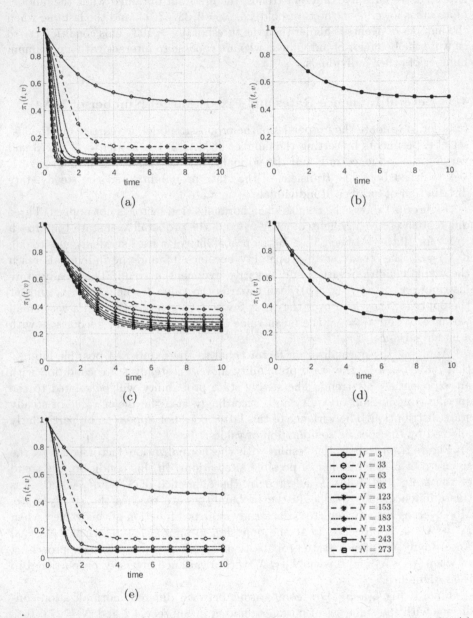

Fig. 4. $\pi_1(t, \mathbf{v})$ when N changes, in: (a) configuration $C1$, (b) configuration $C2$, (c) configuration $C3$, (d) configuration $C4$ and $Th = 10$, (e) configuration $C4$ and $Th = 100$

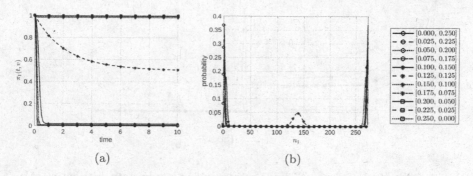

Fig. 5. (a) $\pi_1(t, \mathbf{v})$ with $N = 273$ in configuration $C1$; (b) steady state distribution of n_1

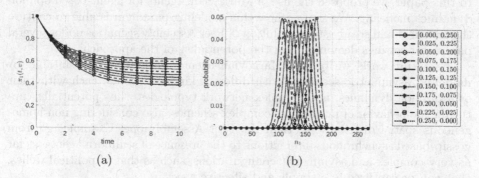

Fig. 6. (a) $\pi_1(t, \mathbf{v})$ with $N = 273$ in configuration $C2$; (b) corresponding steady state distribution of n_1

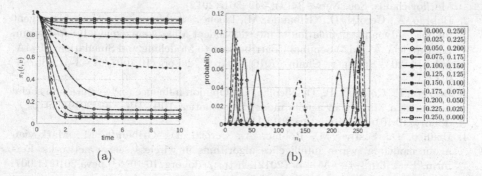

Fig. 7. (a) $\pi_1(t, \mathbf{v})$ with $N = 273$ in configuration $C3$; (b) corresponding steady state distribution of n_1

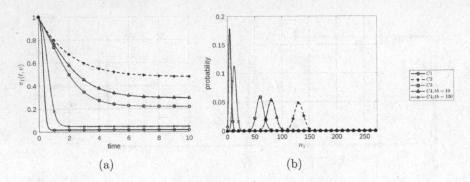

Fig. 8. (a) $\pi_1(t, \mathbf{v})$ with $N = 273$; (b) corresponding steady state distribution of n_1

5 Conclusions and Future Work

In this paper, we proposed the use of Markovian Agents for agent-based opinion dynamics modeling. We showed how this modeling paradigm is able to capture the transient and steady-state evolution of Peer Assembly scenarios under several parameter settings demonstrating the potentiality of the approach.

As future work, we believe Markovian agents are particularly suitable for modeling asymmetric scenarios with different kinds of agents, each with their own opinion dynamics, and complex network typologies, thus potentially capturing the behavior of much more complex schemes, also considering non homogeneous spatial distributions, than the Peer Assembly one. Examples go from gossip-based asynchronous interactions to the presence of stubborn agents, as far as very complex and asymmetric configurations, such as that of political rallies, that can be captured in a simple and effective way.

References

1. Banisch, S., Lima, R., Araújo, T.: Agent based models and opinion dynamics as Markov chains. Soc. Netw. **34**(4), 549–561 (2012)
2. Bobbio, A., Cerotti, D., Gribaudo, M., Iacono, M., Manini, D.: Markovian agent models: a dynamic population of interdependent Markovian agents. In: Al-Begain, K., Bargiela, A. (eds.) Seminal Contributions to Modelling and Simulation. SFMA, pp. 185–203. Springer, Cham (2016). https://doi.org/10.1007/978-3-319-33786-9_13
3. Bolzern, P., Colaneri, P., De Nicolao, G.: Opinion influence and evolution in social networks: a Markovian agents model. Automatica **100**, 219–230 (2019). https://doi.org/10.1016/j.automatica.2018.11.023
4. Bruneo, D., Scarpa, M., Bobbio, A., Cerotti, D., Gribaudo, M.: Markovian agent modeling swarm intelligence algorithms in wireless sensor networks. Perform. Eval. **69**(3–4), 135–149 (2012). https://doi.org/10.1016/j.peva.2010.11.007. Selected papers from ValueTools 2009
5. El-Diraby, T., Shalaby, A., Hosseini, M.: Linking social, semantic and sentiment analyses to support modeling transit customers' satisfaction: towards formal study of opinion dynamics. Sustain. Cities Soc. **49**, 101578 (2019)

6. Gribaudo, M., Cerotti, D., Bobbio, A.: Analysis of on-off policies in sensor networks using interacting Markovian agents. In: 2008 Sixth Annual IEEE International Conference on Pervasive Computing and Communications (PerCom), pp. 300–305. IEEE (2008). https://doi.org/10.1109/PERCOM.2008.100

7. Mastroeni, L., Vellucci, P., Naldi, M.: Agent-based models for opinion formation: a bibliographic survey. IEEE Access **7**, 58836–58848 (2019)

8. Proskurnikov, A.V., Tempo, R.: A tutorial on modeling and analysis of dynamic social networks. Part I. Annu. Rev. Control. **43**, 65–79 (2017)

9. Scarpa, M., Molica, G.: MAGNET: a software library for Markovian agent networks. In: Proceedings of the 11th EAI International Conference on Performance Evaluation Methodologies and Tools, pp. 164–169. Association for Computing Machinery, New York (2017). https://doi.org/10.1145/3150928.3150955

10. Suarez-Lledo, V., Alvarez-Galvez, J., et al.: Prevalence of health misinformation on social media: systematic review. J. Med. Internet Res. **23**(1), e17187 (2021)

11. Urena, R., Kou, G., Dong, Y., Chiclana, F., Herrera-Viedma, E.: A review on trust propagation and opinion dynamics in social networks and group decision making frameworks. Inf. Sci. **478**, 461–475 (2019)

12. Zha, Q., et al.: Opinion dynamics in finance and business: a literature review and research opportunities. Financ. Innov. **6**(1), 44 (2020). https://doi.org/10.1186/s40854-020-00211-3

Using the ORIS Tool and the SIRIO Library for Model-Driven Engineering of Quantitative Analytics

Laura Carnevali[1] , Marco Paolieri[2] , Riccardo Reali[1][(✉)] ,
Leonardo Scommegna[1] , Federico Tammaro[1], and Enrico Vicario[1]

[1] Department of Information Engineering, University of Florence, Florence, Italy
{laura.carnevali,riccardo.reali,leonardo.scommegna,federico.tammaro,
enrico.vicario}@unifi.it
[2] Department of Computer Science, University of Southern California,
Los Angeles, USA
paolieri@usc.edu

Abstract. We present a Model-Driven Engineering (MDE) approach to quantitative evaluation of stochastic models through the ORIS tool and the SIRIO library. As an example, the approach is applied to the case of a tramway line with reduced number of passengers to contain the spread of infection during a pandemic. Specifically, we provide a meta-model for this scenario, where, at each stop, only a certain number of people can ride the tram depending on the current tram capacity, the length of the queue of people waiting at the stop, and the number of passengers on the tram. Then, the ORIS tool and the SIRIO library are used as a software platform to derive a Stochastic Time Petri Net (STPN) representation for each tramway stop and to perform its regenerative transient analysis to obtain quantitative measures of interest, such as the expected number of people waiting at each stop and the expected number of tram passengers over time. Experimental results show that the approach facilitates exploration of the space of design choices, providing insight about the effects of parameter changes on quantitative measures of interest and allowing balanced queue sizes at different stops.

Keywords: Quantitative evaluation · Model Driven Engineering (MDE) · Software tools and libraries · Intelligent transportation systems

1 Introduction

Models are generally used to replace the system under study with a domain-focused although simplified view [5]. By shifting the attention primarily to models, the concepts expressed are less bound to a precise technology or framework, and closer to the problem domain [13], allowing better insight into the issues of interest: this approach eases system understanding by domain experts, improves

K. Gilly and N. Thomas (Eds.): EPEW 2022, LNCS 13659, pp. 200–215, 2023.
https://doi.org/10.1007/978-3-031-25049-1_13

the expressivity of the system description, and facilitates the maintainability of the adopted solution. At the same time, models also support early evaluation of the impact of design choices on the final behavior before data become available, enabling fast exploration of the space of possible solutions.

Model-Driven Engineering (MDE) is an approach that considers models as primary artifacts during all software development phases [5] and connects them with the practice of software engineering, making them living components of the system rather than using them for documentation and study purposes only. MDE usually consists in the use of Domain-Specific Languages (DSLs) and Domain-Specific Modeling Languages (DSMLs), specialized in formalizing the structure and behavior of applications, and described using meta-models to map relationships, semantics, and constraints between concepts expressed in a domain [12]. This practice is also common in industrial contexts, where small DSLs are developed for narrow and well-understood domains [16]. Another important aspect that led to the widespread use of MDE and Model-Driven Design (MDD) are automated transformations: Model-to-Text (M2T) transformations are more commonly used to transform a particular model instance into text-based file formats or software artifacts available as source code (code generation), while Model-to-Model (M2M) transformations are used to translate a model into another model. Both techniques are referred to as "correct-by-construction" [12] given that they do not require any subsequent modification and they avoid manual, and thus error-prone, changes to the considered artifacts.

In a broader perspective, models can be generated at system runtime with an inverse approach, i.e., by connecting models to operational data, possibly persisted in a data layer, and by enabling the management of high volumes of concrete running instances derived from a meta-model, with variability determined by runtime changes (e.g., time-dependent parameters). When combined with the use of formal semantics and solution techniques to compute results of a service request, this two-way approach supports dynamic state monitoring and system control during execution as well as understanding of runtime behavior, including the identification of behavioral phenomena [1]. In so doing, the approach allows agile validation of choices in the design phase.

In this paper, we outline an MDE approach which demonstrates how to leverage the ORIS suite [11,14] to develop practical applications of stochastic modeling and analysis. In particular, we focus on how to manage crowding of a tramway line to contain the spread of a pandemic, which comprises a problem of resource assignment: at each stop, a transit pass can be granted only to a certain number of people, depending on the current capacity of the tram, the length of the queue of people waiting at the stop, and the number of passengers on the tram. To this end, we perform a context analysis to derive a meta-model of the considered scenario. Then, we exploit MDE practices and M2M transformations to derive a Stochastic Time Petri Net (STPN) representation for each tramway stop, and then we perform regenerative transient analysis [8] of each tramway stop model. Specifically, the analysis of each stop provides the expected number of queued people over time and the expected number of tram passengers over time, showing that the distribution of the number of queued people and the distribution of the number of tram passengers at the beginning of each tram period reach a steady state within a limited number of tram periods. Then, the steady state distribution of the number of tram passengers at the beginning of tram periods is used in the analysis of the subsequent tramway stop as the initial distribution of the number of tram passengers. In the theoretical perspective, the tramway stop model fits in the class of polling systems [9], where the periodic arrival pattern of trams is impacted by a stochastic delay (jitter). Experimental results shows that the approach facilitates exploration of the design space, providing insight about the effects of parameter changes on performance measures of interest, such as the expected number of queued people at each stop and the expected number of passengers on the tram.

The rest of this paper is organized as follows. In Sect. 2, we describe the application domain and we provide a meta-model to define context-specific models. In Sect. 3, we illustrate how the ORIS tool supports MDE of quantitative analytics, transforming a tramway line model into an STPN and performing its evaluation through regenerative transient analysis. In Sect. 4, we explain how to support modeling and evaluation steps through the SIRIO library. In Sect. 5, we illustrate the experimental results obtained analyzing the models of a tramway line. Finally, conclusions are drawn in Sect. 6.

2   A Tramway Line Meta-model

Figure 1 shows a meta-model of the tramway scenario, designed to express a wide range of models at different levels of granularity and derive (through M2M transformations) analytic representations to get insight into the scenario. Specifically, the meta-model represents a network of tramway lines (represented by the Network and Route classes) each characterized by a TimeBand identifying the time period of tramway service and the frequency of tram departures. An itinerary is an ordered sequence of StopPoint instances, each characterized by a StopPointCapacity which can be affected by different *restriction policies*, e.g., due to maintenance works or social distancing. Each Tram has a maximum *capacity* in terms of number of passengers (possibly subject to restriction policies), runs on a specific route, and stops at each stop point of the route itinerary.

The Stop class identifies the event that a tram of a certain route arrives and stops at a specific stop point. It is described by: an ArrivalTimeTable instance, characterized by a *jitter* delay with respect to the nominal arrival time; a maximum number of passengers that can board the tram (with the possibility to define an ad-hoc restriction policy); a flow of *passengers* of different *types*. The latter association enables the representation of a stop point visited by a heterogeneous group of passengers with a different arrival rate, outgoing rate (i.e., the tendency to abandon the stop point) and disembark rate. The stop event is characterised by parameters that may vary with the time of day (e.g., a tram stop near a school is characterized by different flows of passengers immediately after school hours and during a night time slot), and thus it is also characterized by a *time chunk*.

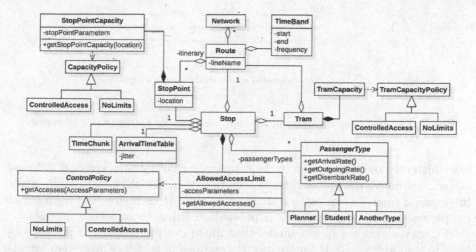

Fig. 1. Meta-model of a network of tramway lines.

Note that the class diagram representing the meta-model could be exploited to define the domain model of a software architecture. In this context, the entire framework could be transposed into a *Software as a Service* (SaaS) system of a cloud infrastructure. On the one hand, the class diagram is designed to enable an easy mapping of object instances to the database through standard Object-Relational Mapping (ORM) technologies (e.g., JPA for Java EE). On the other hand, the M2M transformation procedure, as well as the quantitative analysis through the SIRIO library, represent business capabilities that could be conveniently encapsulated into a set of microservices, generating a cloud architecture able to fulfill various use cases such as those illustrated in Fig. 2. In particular, to avoid crowding on the tram and at tramway stops, the mentioned M2M transformations (see Sects. 3 and 4) enable the evaluation of the expected value of queued people at a stop and of people traveling on a tram. Such information could be exploited by a tramway schedule planner to modulate the frequency of trams during a certain time band, to change the maximum capacity of a tram, or, more generally, to adapt system parameters to guarantee some safety conditions.

3   A Tramway Stop Model

We define a model of a tram stop (Sect. 3.1), and we use the ORIS tool to derive its STPN representation (Sect. 3.2) and to compute rewards of interest through regenerative transient analysis of the STPN model (Sect. 3.3).

3.1   Model Description

We consider a model in the class of polling systems [1], i.e., systems where queue processes are served by a component, which does not operate continuously, but

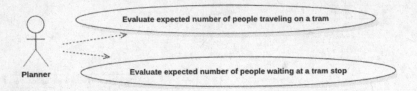

Fig. 2. Use case diagram showing relevant use cases for tramway line planning.

intermittently and recurrently. In the tramway scenario, trams arrive at different stops according to a probability distribution that combines the deterministic interarrival time of trams with a stochastic jitter. At each stop, the tram serves people waiting in a queue that is populated according to an arrival process.

Figure 3 shows the object model that illustrates the considered polling system, in conformity with the meta-model described in Sect. 2. The system targets a single *stop point* (SP-A) of a tramway line (R1). To avoid crowding, a stop point is associated with a *capacity* (CPT-A) characterized by a rejection policy, which disables queue access to people arriving after the saturation of maximum queue capacity, here equal to 10. A route is characterized by a *frequency*, which is the inverse of the deterministic time between the arrival of two consecutive trams at the stop point. Each individual *stop* event (SA1) is also characterized by a stochastic delay defined by a *jitter*. In our model, we assume an interarrival time equal to 220s seconds and a jitter distributed as an expolynomial function $f(x) = 0.075468 \exp(-x/10) + 0.002516\,x \exp(-x/10)$ with support $[0, 60]$ s. Different classes of people (STD-P) may arrive at the stop point with a specific probability distribution; for simplicity, we only consider *standard passengers* arriving at the queue according to a Poisson process with rate 0.08 passengers/second. Finally, we consider trams with maximum capacity (CPT-T) equal to 4; additional people waiting to board the tram are rejected (CA-S control policy).

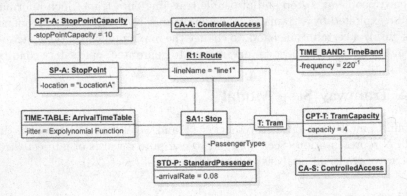

Fig. 3. Object model of a tram stop.

3.2 STPN Representation

Figure 4 shows the STPN model of a single tram stop. In an STPN, time processes are modeled with transitions, represented as bars with different colors depending on the probability density function (PDF) of their associated timers: transitions with exponential PDFs (EXP) are represented as white thick bars (e.g., transition passengerArrival); transitions with deterministic times (DET) are represented as gray thick bars (e.g., transition serviceArrivalNominal); transitions with non-exponential general distributions (GEN) are represented as black thick bars (e.g., transition serviceArrival). A deterministic transition with time 0 is called *immediate* (IMM) and it is represented as a black thin bar (e.g., transitions leaving and boarding). Discrete logical states of the system are modeled as tokens within places, which are represented as dots or numbers within circles (e.g., places TramCapacity and QueueCapacity, respectively). Places and transitions are connected with directed arcs, modeling precedence relations among processes. Directed arcs from input places to transitions define preconditions, while arcs from transitions to output places defines postconditions (e.g., the tuple (WaitJitterDelay, serviceArrival) is a precondition).

A transition becomes *enabled* by a marking (i.e., an assignment of tokens to one or more places) when each of the input places contains at least one token (e.g., if a token is added to place WaitJitterDelay, transition serviceArrival is enabled), and when an *enabling function* (if any) is satisfied for the transition (e.g., transition passengerArrival is enabled when the tokens in place Queue are less than or equal to those in place QueueCapacity). For each enabled transition t, a time-to-fire is sampled between its earliest firing time (EFT) and latest firing time (LFT), according to the PDF $f_t(x)$ of the transition. When its time-to-fire elapses, a transition becomes *firable*; when it fires, one token is removed from each of its input places and one token is added to each of its output places (e.g., if transition passengerArrival fires, then a token is moved from place WaitJitterDelay to place TrainArrived), or the marking is modified according to an *update function* (e.g., the firing of transition boarding removes all tokens from place Queue, when the queue is less than the tram capacity, or subtracts from place Queue a number of tokens equal to the remaining capacity of the tram, when the queue is larger). In case of multiple firable transitions, transitions with lower *priority number* have precedence; among transitions with the same priority, one is selected with a random switch where probabilities are

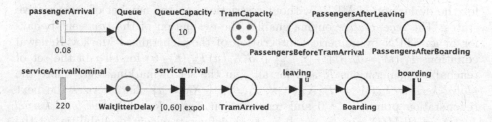

Fig. 4. The STPN of a tram stop.

proportional to *transitions weights*. After a firing, the times-to-fire of transitions enabled before and after the token moves are reduced by the time-to-fire of the fired transition (e.g., if `passengerArrival` fires before `serviceArrival`, the time-to-fire of the latter transition is decreased by the time elapsed since the previous firing).

The STPN model of Fig. 4 can be evaluated to get insight about the model described in Fig. 3. Transitions `serviceArrivalNominal` and `serviceArrival` model the nominal time and the stochastic jitter employed by the tram to periodically reach the considered stop. Transition `passengerArrivial` models the arrival process of people at the tram stop, which is a Poisson process, as described in Sect. 3.1. Finally, IMM transitions `leaving` and `boarding` are introduced to represent the processes of exiting and entering the tram, which are assumed to happen in zero time. Note that, in the model of Fig. 4, places are used not only to represent logical states of a specific process, but also to represent model parameters (e.g., `QueueCapacity` and `TramCapacity`) or variable counters (e.g., `Queue`, `PassengersBeforeTramArrival`, `PassengersAfterLeaving` and `PassengersAfterBoarding`).

3.3   Model Evaluation and Rewards of Interest

STPN models enable the application of analysis techniques that can produce insight, analytics or predictions on such models. The ORIS tool provides a GUI to draw system models and to apply such analysis methods. Evaluation of a model involves identifying certain quantities of interest, which can be formulated as *rewards*. In ORIS, a reward is an expression including constants and token counts, which defines a real-valued function over markings of the STPN. A reward is used to specify quantitative measures of interest to be evaluated by the analysis engines of ORIS; for example, to evaluate the evolution over time of the number of queued people in the model of Fig. 4, it is sufficient to evaluate the reward "`Queue`".

The ORIS tool provides different analysis engines [11] enabling rewards evaluation through different techniques. The *nondeterministic* engine is designed to enable nondeterministic analysis of the state space of an STPN; the *Markovian* engine implements methods to evaluate models with underlying Continuous Time Markov Chains (CTMC); the *enabling restriction* engine is tailored for the evaluation of Markov Regenerative Processes (MRPs) under the enabling restriction [7]; the *forward* and the *regenerative* engines implement different techniques for the evaluation of MRPs without restrictions on the number of regenerative states. The *regenerative* engine enables the evaluation of the transient behavior of an STPN by numerical integration of the Generalized Markov Renewal equations $P_{ij}(t) = L_{ij}(t) + \sum_{k \in \mathcal{R}} \int_0^t dG_{ik}(u) P_{kj}(t-u)$ for all i in the set of reachable regenerations $\mathcal{R}$ and for all j in the set of markings $\mathcal{M}$, where the *global kernel* $G_{ik}(t) := P\{X_1 = k, T_1 \le t \mid X_0 = i\}$ characterizes the next regeneration point $T_1 \ge 0$ and regeneration $X_1 \in \mathcal{R}$, while the *local kernel* $L_{ij}(t) := P\{M(t) = j, T_1 > t \mid X_0 = i\}$ defines transient probabilities of the process until the next regeneration point. Kernels can be evaluated using the method of *stochastic state classes* [8]; this method encodes the marking, the support of the times-to-fire of enabled transitions after each sequence of firings between any two regeneration points, and the continuous joint PDF with a piecewise representation over Difference Bounds Matrix (DBM) zones [3], that can be efficiently evaluated in closed-form by the ORIS tool when each transition has expolynomial PDF [15].

As discussed in Sect. 2, we are interested in two use cases: predicting the expected number of people waiting at a stop and predicting the expected number of people traveling on a tram. According to the semantics of rewards in ORIS, these metrics can be evaluated for the model of Fig. 4 by using the rewards "Queue" and "PassengersAfterBoarding", respectively. The evaluation can be carried out with the GUI of ORIS, or through its analysis library SIRIO, as presented in Sect. 4.

4   Combining SIRIO with MDE Practices

SIRIO is a Java library from the ORIS tool suite which collects modeling, analysis and simulation features for STPNs. In this section, we briefly illustrate how to use the SIRIO library to model and analyze the polling system presented in Sect. 3 (Sect. 4.1) and we describe how MDE practices can transform models from the specification of the meta-model provided in Sect. 2 to the SIRIO API (Sect. 4.2) to evaluate quantitative measures of interest.

4.1   SIRIO Modeling and Evaluation

Figure 5 shows the UML class diagram of the SIRIO STPN modeling component. An STPN can be modeled by instancing a PetriNet object as an aggregation of instances of Place and Transition. Precedence constraints between places and transitions can be obtained by adding instances of Precondition and Postcondition to the PetriNet object. Precondition and Postcondition classes model precedence constraints between places and transitions, and transitions and places, respectively. TransitionFeature instances can be added to a transition: a transition can have an EnablingFunction, a PostUpdater, and a StochasticTransitionFunction, which represent the condition to enable a transition, the policy to update the marking after the transition fires, and the PDF of the time-to-fire of the transition, respectively.

In Listing 1, we provide the SIRIO implementation of the model of the polling system, described in Sect. 3. Initially, a PetriNet and a Marking are created (lines 1 and 2). Then, places are added to the PetriNet through the method addPlace() (lines 3 to 11). The method returns a Place object, which is used to reference the newly created place to add preconditions and postconditions. Similarly, transitions are created through the method addTransition() (lines 12 to 16), which is called on the PetriNet object and returns a Transition object. After creating places and transitions, it is possible to add Precondition and Postcondition instances through the methods addPrecondition() and

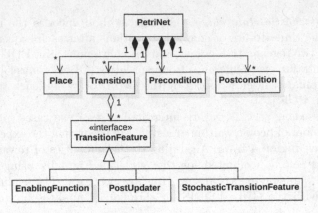

Fig. 5. Class diagram of the SIRIO STPN modeling component.

addPostçondition(), respectively (lines 17 to 23). To set the initial marking of the STPN, the method setTokens() can be called for each place (lines 24 to 32). Finally, additional features can be added to transitions through the method addFeature() (lines 33 to 50). Objects of different types can be passed to this method to add different features: EnablingFunction objects (line 42) to add enabling functions, PostUpdater objects (lines 33–35 and 38–39) to add update functions, and StochasticTransitionFeature objects to add stochastic behavior. Note that the StochasticTransitionFeature class implements static constructors to facilitate the creation of the different stochastic transition types presented in Sect. 3.2: newDeterministicInstance() for DET transitions (lines 48–49) and IMM transitions (lines 36 and 40); newExponentialInstance() (lines 43–44) for EXP transitions; and newExpolynomialInstance() (lines 45–47) for expolynomial transitions.

After the creation of the model in the SIRIO specification, regenerative transient analysis can be performed. In Listing 2, analysis time limit, time tick and error threshold are set (lines 1 to 3); these parameters are used to create a builder of a RegTransient object (lines 6 to 10), which is then instantiated and used to perform the analysis (lines 12 to 14); finally, the reward Queue (or PassengersAfterBoarding) is evaluated on the solution (lines 17 and 18).

4.2   Mapping Meta-models to SIRIO Petri Nets

Model-to-model (M2M) transformations are used to move from one domain to another one much closer to the solution domain [5]. In our case study, the transformation between the proposed meta-model and the Petri net classes used by SIRIO is automated by following a set of mapping rules for the entire process. While there is loss of information in the transition from the source model to the target model, STPNs are used only for analysis purposes and all the required information is present in the solution domain. Following the classification in [4], this M2M process can be seen as a unidirectional transformation that computes a target model from a source model; it uses a relational approach with a set of mapping rules to link source and target element types.

```
1   PetriNet net = new PetriNet();
2   Marking marking = new Marking();
3   Place Boarding = net.addPlace("Boarding");
4   Place WaitJitterDelay = net.addPlace("WaitJitterDelay");
5   Place TramArrived = net.addPlace("TramArrived");
6   Place TramCapacity = net.addPlace("TramCapacity");
7   Place Queue = net.addPlace("Queue");
8   Place QueueCapacity = net.addPlace("QueueCapacity");
9   Place PassengersAfterBoarding = net.addPlace("PassengersAfterBoarding");
10  Place PassengersAfterLeaving = net.addPlace("PassengersAfterLeaving");
11  Place TramPassengersBeforeTramArrival = net.addPlace("TramPassengersBeforeTramArrival");
12  Transition serviceArrivalNominal = net.addTransition("serviceArrivalNominal");
13  Transition serviceArrival = net.addTransition("serviceArrival");
14  Transition leaving = net.addTransition("leaving");
15  Transition boarding = net.addTransition("boarding");
16  Transition passengerArrival = net.addTransition("passengerArrival");
17  net.addPostcondition(serviceArrivalNominal, WaitJitterDelay);
18  net.addPrecondition(WaitJitterDelay, serviceArrival);
19  net.addPostcondition(serviceArrival, TramArrived);
20  net.addPrecondition(TramArrived, leaving);
21  net.addPostcondition(leaving, Boarding);
22  net.addPrecondition(Boarding, boarding);
23  net.addPostcondition(passengerArrival, Queue);
24  marking.setTokens(WaitJitterDelay, 1);
25  marking.setTokens(TramArrived, 0);
26  marking.setTokens(Boarding, 0);
27  marking.setTokens(Queue, 0);
28  marking.setTokens(QueueCapacity, 10);
29  marking.setTokens(TramCapacity, 4);
30  marking.setTokens(TramPassengersBeforeTramArrival, 0);
31  marking.setTokens(PassengersAfterLeaving, 0);
32  marking.setTokens(PassengersAfterBoarding, 0);
33  boarding.addFeature(new PostUpdater(String.join(
34      "PassengersAfterBoarding=min(PassengersAfterLeaving+Queue,TramCapacity);",
35      "Queue=max(0, Queue-TramCapacity+PassengersAfterLeaving);"), net));
36  boarding.addFeature(StochasticTransitionFeature.newDeterministicInstance("0"));
37  boarding.addFeature(new Priority(0));
38  leaving.addFeature(new PostUpdater(
39      "PassengersAfterLeaving=max(0,PassengersBeforeTramArrival-1);", net));
40  leaving.addFeature(StochasticTransitionFeature.newDeterministicInstance("0"));
41  leaving.addFeature(new Priority(0));
42  passengerArrival.addFeature(new EnablingFunction("Queue<QueueCapacity"));
43  passengerArrival.addFeature(StochasticTransitionFeature.newExponentialInstance(
44      configuration.rate()));
45  serviceArrival.addFeature(StochasticTransitionFeature.newExpolynomial(
46      "0.075467664 * Exp[-0.1 x] + 0.0025155888 * x^1 * Exp[-0.1 x]",
47      new OmegaBigDecimal("0"), new OmegaBigDecimal("60")));
48  serviceArrivalNominal.addFeature(StochasticTransitionFeature.newDeterministicInstance(
49      configuration.serviceArrivalNominalTime()));
50  serviceArrivalNominal.addFeature(new Priority(0));
```

Listing 1: SIRIO implementation of the model described in Fig. 4.

```
1    BigDecimal bound = new BigDecimal("2200.0");
2    BigDecimal step = new BigDecimal("10.0");
3    BigDecimal epsilon = new BigDecimal("0.001");
4
5    // analyze
6    RegTransient.Builder builder = RegTransient.builder();
7    builder.timeBound(bound);
8    builder.timeStep(step);
9    builder.greedyPolicy(bound, epsilon);
10   builder.markingFilter(MarkingCondition.fromString("Queue"));
11
12   RegTransient analysis = builder.build();
13   TransientSolution<DeterministicEnablingState, Marking> probs =
14           analysis.compute(net, marking);
15
16   // evaluate reward
17   TransientSolution<DeterministicEnablingState, RewardRate> reward =
18           TransientSolution.computeRewards(false, probs,
19                                    RewardRate.fromString("Queue"));
```

Listing 2: SIRIO implementation of the regenerative transient analysis required to evaluate the reward Queue.

An instance of the meta-model referring to a precise train network can be created from JSON input, or retrieved from the database and scanned with a dedicated *Visitor* [6] to analyze its structure: the Visitor retrieves all the required information associated to a precise Route instance, then each stop is treated separately by creating different PetriNet objects and analyzed independently. Each model is constructed following the structure introduced in Fig. 4 and then populated with the parameters that characterize a particular stop:

- The rate of the EXP transition passengerArrival is automatically calculated as the sum of incoming rates of all passenger types;
- TramCapacity and QueueCapacity places are populated according the access policy of the stop and tram instances (with controlled access or no limits);
- the PDF of transitions serviceArrivalNominal and serviceArrival is obtained from the frequency and jitter attributes of the tram stop.

This procedure completely automates the transformation between models and is implemented through the programming API and classes provided by the SIRIO library. This approach avoids manual and error-prone construction of a model for each stop of a tram network.

Table 1. Queue capacity and arrival rate (passengers/s) of each stop.

Parameter	Stop							
	1	2	3	4	5	6	7	8
Queue capacity	25	50	25	50	25	50	25	50
Arrival rate	0.01	0.01	0.02	0.02	0.04	0.04	0.08	0.08

5   Experimentation

In this section, we use the SIRIO library to evaluate quantitative metrics for the case of an 8-stop tramway line. In particular, we describe our setup (Sect. 5.1) and then we discuss the obtained results (Sect. 5.2).

5.1   Experimentation Setup

We consider the case of a tramway with 8 stops, each modeled as the system of Fig. 4 with a tram interarrival time equal to 220 s, but with different queue capacities and passenger arrival rates. To reduce the complexity of the analysis, we consider the arrival process of people at a stop, and the boarding and the leaving processes on/from the tram as batch processes. In particular, a token in places Queue, QueueCapacity, TramCapacity, PassengersBeforeTrain Arrival, PassengersAfterLeaving and PassengersAfterBoarding represents a group of exactly 5 people. In so doing, the tram capacity is set to 20 people (4 tokens), while queue capacity depends on the considered stop, and can have 25 or 50 people (5 or 10 tokens). Different stops are analyzed with SIRIO in sequence, using the steady state number of people that already are on the tram obtained from the analysis of the previous stop. To this end, we evaluate the expected steady state value of the rewards PassengersAfterBoarding==0, PassengersAfterBoarding==1, PassengersAfterBoarding==2, PassengersAf terBoarding==3, and (for a full tram) PassengersAfterBoarding==4. The obtained probabilities are used as the weights of a random switch that draws the number of people on the tram arriving at the next stop, i.e., which updates the number of tokens in place PassengersBeforeTrainArrival with a number of tokens between 0 and 4 (0, 5, 10, 15, or 20 people). Finally, we assume that, when the tram arrives, exactly 5 people leave it (1 token); then, as many waiting people as possible board the tram.

The analysis of a stop is performed by evaluating transient rewards Queue and PassengersAfterBoarding using regenerative analysis based on the method of stochastic state classes [8], with a time limit equal to 2200 s. Parameters of models of different tram stops (reported in Table 1) are selected with the intent of evaluating how changes can influence these metrics.

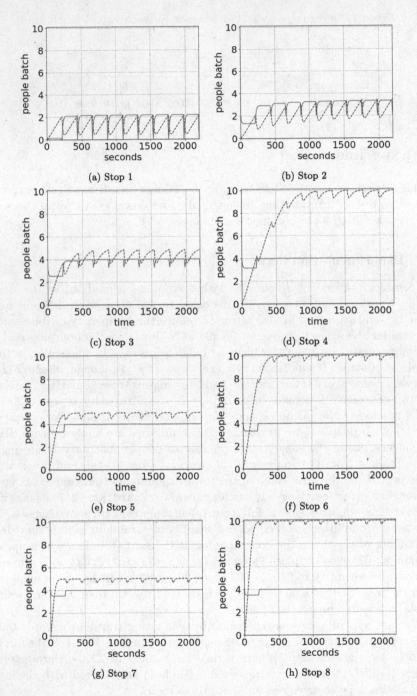

Fig. 6. Expected number of groups of 5 people waiting for a tram at the stop (blue dashed line) and traveling on a tram (red solid line), for each stop, as a function of time (in s). (Color figure online)

5.2   Experimentation Results

The graphs in Fig. 6 show the transient behaviors of the expected number of queued people at the stop (blue dashed line) and people traveling on a tram (red solid line), for each stop. In Sect. 5.2 and Sect. 5.2, the arrival rate at the stop is equal to 0.01 passengers/s, and it can be observed that the queue does not saturate or saturates very slowly, respectively. In particular, since empty trams periodically arrive at the stop, in Sect. 5.2 the queue is always emptied before reaching its capacity. Moreover, the expected number of people in the queue at the beginning of each tram period reaches a steady state. As the rate increases (Sect. 5.2 to Sect. 5.2), queue saturation occurs earlier. This can be mitigated by increasing the queue capacity (compare, for example, Sect. 5.2 and Sect. 5.2); however, already with arrival rate equal to 0.04 s (Sect. 5.2 and Sect. 5.2), the queue saturates within the first two periods of tram arrival.

The red curve in each figure shows an initial period in which queued people wait for the first tram to pass. With the exception of the first stop, where the first tram arrives empty (the red line starts at 0), at the other stops, the tram arrives with a number of passengers that depends on the number of people who left the previous stop. After the first period, the distribution of the expected number of people on the tram exhibits a periodic behavior. Since it is assumed that as many passengers as possible board the tram, as soon as the queue saturates, the tram becomes fully occupied, and the expected number of people on the tram reaches its maximum value. In addition, this behavior occurs very quickly in the final stops, which happens because trams arrive at the final stops already full. For example, since stop 3 tram becomes full within the third period, the next stop distributions reach the steady state almost immediately.

6   Conclusions

In this work, we illustrated how to use the ORIS tool and the SIRIO library as a software platform to develop quantitative predictive analytics. To this end, we combined a practical perspective, aimed at avoiding crowding at tramway stops to limit the spread of infection in a pandemic, with a formalized abstraction in the framework of the theory of polling systems, highlighting how the two perspectives are effectively connected through practices of MDE. Note that, while the scenario of tramway lines is used as an application example to show that ORIS and SIRIO are easily usable and provide effective support to develop efficient analytics, stochastic models for distributed concurrent systems have been widely used to represent crowd scenarios, typically exploiting fluid flow analysis based on the solution of ordinary differential equations. As an example, in [2], stochastic process algebra and stability analysis are used to evaluate coordination of agents in large collective systems. Stochastic process algebra and fluid flow approximation are used also in [10] to study emergent crowd behavior.

Experimental results achieved in Sect. 5 suggest that, though the expected number of queued people at a tram stop is recurrently perturbed by the process of tram arrivals and the expected number of tram passengers is recurrently perturbed by the process of leaving and boarding at the tram stops, the distribution of the number of queued people and the distribution of number of tram passengers at the beginning of each tram period actually reach a steady state, in just a few periods. Therefore, the steady state distribution of the number of tram passengers computed at a tram stop can be considered as the initial distribution of the number of tram passengers at the subsequent tram stop, enabling the formulation of an optimization problem where the system parameters are selected so as to balance the queue size at the various stops along the line.

Future work includes defining a theoretical framework to characterize the obtained experimental results, extending the model with additional parameters such as different limits on the number of people that can ride the tram at each stop of a line, and performing a broader experimentation by varying a larger set of parameters for a larger number of parameter values.

References

1. Blair, G., Bencomo, N., France, R.B.: Models@ run.time. Computer **42**(10), 22–27 (2009)
2. Bortolussi, L., Latella, D., Massink, M.: Stochastic process algebra and stability analysis of collective systems. In: De Nicola, R., Julien, C. (eds.) COORDINATION 2013. LNCS, vol. 7890, pp. 1–15. Springer, Heidelberg (2013). https://doi.org/10.1007/978-3-642-38493-6_1
3. Carnevali, L., Grassi, L., Vicario, E.: State-density functions over DBM domains in the analysis of non-Markovian models. IEEE Trans. Softw. Eng. **35**(2), 178–194 (2009)
4. Czarnecki, K., Helsen, S.: Classification of model transformation approaches. In: Proceedings of the 2nd OOPSLA Workshop on Generative Techniques in the Context of the Model Driven Architecture, vol. 45, pp. 1–17 (2003)
5. Da Silva, A.R.: Model-driven engineering: a survey supported by the unified conceptual model. Comput. Lang. Syst. Struct. **43**, 139–155 (2015)
6. Gamma, E., Helm, R., Johnson, R.E., Vlissides, J.: Design Patterns: Elements of Reusable Object-Oriented Software. Addison-Wesley Professional, Boston (1995)
7. German, R., Logothetis, D., Trivedi, K.S.: Transient analysis of Markov regenerative stochastic Petri nets: a comparison of approaches. In: Proceedings 6th International Workshop on Petri Nets and Performance Models, pp. 103–112. IEEE (1995)
8. Horváth, A., Paolieri, M., Ridi, L., Vicario, E.: Transient analysis of non-Markovian models using stochastic state classes. Perform. Eval. **69**(7–8), 315–335 (2012)
9. Ibe, O.C., Trivedi, K.S.: Stochastic Petri net models of polling systems. IEEE J. Sel. Areas Commun. **8**(9), 1649–1657 (1990)
10. Massink, M., Latella, D., Bracciali, A., Hillston, J.: A combined process algebraic, agent and fluid flow approach to emergent crowd behaviour. Tech. rep., CNR-ISTI (2010)
11. Paolieri, M., Biagi, M., Carnevali, L., Vicario, E.: The ORIS tool: quantitative evaluation of non-Markovian systems. IEEE Trans. Softw. Eng. **47**(6), 1211–1225 (2019)
12. Schmidt, D.C.: Model-driven engineering. Comput.-IEEE Comput. Soc. **39**(2), 25 (2006)

13. Selic, B.: The pragmatics of model-driven development. IEEE Softw. **20**(5), 19–25 (2003). https://doi.org/10.1109/MS.2003.1231146
14. Sirio: the Sirio library for the analysis of stochastic time Petri nets. https://github.com/oris-tool/sirio
15. Trivedi, K.S., Sahner, R.: SHARPE at the age of twenty two. ACM SIGMETRICS Perform. Eval. Rev. **36**(4), 52–57 (2009)
16. Whittle, J., Hutchinson, J., Rouncefield, M.: The state of practice in model-driven engineering. IEEE Softw. **31**(3), 79–85 (2013)

Places, Transitions and Queues: New Proposals for Interconnection Semantics

Marco Gribaudo[1] and Mauro Iacono[2(✉)]

[1] Politecnico di Milano, via Ponzio 4/5, 20133 Milan, Italy
marco.gribaudo@polimi.it
[2] Università degli Studi della Campania "L. Vanvitelli",
viale Lincoln 5, 81100 Caserta, Italy
mauro.iacono@unicampania.it

Abstract. In this paper we discuss some novel considerations about the semantics of multiformalism models in which Petri Nets (PN) and Queuing Networks (QN) are coupled. These considerations aim to further increase the modeling power of the interconnections between places and transitions on one side and queues on the other. Although this type of interconnection has been previously addressed in the literature, during the works to extend the Java Modeling Tools (JMT) for adding PN primitives to existing QN models, we found that a wider range of interpretations of the interconnections between the two formalisms is possible, including cases that cannot be immediately tracked back to states superposition and events synchronization, but that can be easily implemented in discrete event simulation. In this work we present and discuss some of the most interesting scenario identified, with their potential applications.

Keywords: Petri nets · Queuing networks · Semantics · Multiformalism modeling

1 Introduction

In this paper we discuss some novel considerations about the semantics of multiformalism models in which Petri Nets (PN) and Queuing Networks (QN) are coupled. These considerations aim to further clarify the possibilities offered by a new analysis of the semantics of interconnection between places and transitions on one side and queues on the other, and have been fostered by activities carried out for the extension in this direction of a well known tool, Java Modeling Tools [6]. In particular, these extensions can help describing complex queuing policies and control strategies, where for example, servers are dynamically allocated, jobs are transferred in bulks to other nodes, or complex synchronization schemes are in place. In fact, previous literature exists on the topic of multiformalism modeling tools capable of implementing the interoperability between PN and QN on the semantic level based approach, including our previous work on the SIMTHE*Sys* modeling approach [12] and the main reference framework

K. Gilly and N. Thomas (Eds.): EPEW 2022, LNCS 13659, pp. 216–230, 2023.
https://doi.org/10.1007/978-3-031-25049-1_14

for the efficient evaluation of multiformalism models, Möbius [8]. The most common interaction semantics interpretation in literature can be connected to the abstract notions of *state* and *event*, on which respectively the characteristics of PN places and queuing sections of QN stations on one side, and PN transitions and service components of QN nodes on the other, may be mapped[1]: this is, by the way, the approach that has been pursued, on different premises, both in our previous work and by Möbius. Interactions between formalisms may be conceived on this abstract level by means of a generalization based on metamodeling or by resorting to states superposition and events synchronization.

Working on JMT, that is based on a different, simulation-oriented approach originally focused on QN, we found that a wider range of interpretations of the interconnections between the two formalisms is possible, including cases that cannot immediately tracked back to states superposition and events synchronization. Conversely, they can actually be implemented in SIMTHE*Sys*, a multiformalism modeling framework designed to allow experimenting with new formalisms or extensions of existing formalisms. In this paper we present and discuss some of the most interesting cases we identified, to point out the original aspects behind them and the perspective of experimenting with them in SIMTHE*Sys* to prepare the implementation in JMT.

This paper is organized as follows: Sect. 2 present related work, Sect. 3 introduces generalities about multiformalism modeling and the case of QN and PN, Sect. 4 proposes a conceptual equivalence between queues and PN suitable to support our proposal, and Sect. 5 provides a running example, followed by conclusions and future work description.

2   Related Work

Literature presents several examples of joint use of PN and QN, in different fashions, to conjugate the advantages and immediateness of the two formalisms or to benefit of their evaluation techniques or expressive power. For example, an early, and illuminating, case is presented in [1], in which the benefits of a combined use of Generalized Stochastic Petri Nets (GSPN) and Product Form Queuing Networks (PFQN) are presented. Benefits are presented in terms of ease of representation of the actual behaviors of computer systems and efficiency in the evaluation process deriving from a modular/hierarchical approach, that also preserves the capability of interpreting intermediate results about subsystems. Another example [15] defines a methodology for combining GSPN and QN in the perspective of exploiting QN to reduce otherwise very large GSPN. PFQN are exploited by means of a detection process that finds submodels and replaces them with equivalents, inspired to [1] and leveraging some of its results. In [14] Deterministic and Stochastic PN are combined with product form QN to evaluate a class of degradable systems by using a hierarchical approach.

---

[1] We are not interested, in this paper, to also consider model transformation based strategies, such as in [7].

The advantages of conjugating the features offered by PN and QN is also the motivation behind hybrid formalisms, such as Queuing Petri Nets (QPN) [3,13] and Petri Nets including Queuing Networks (PNiQ) [4]. QPN extend Colored GSPN (CGSPN) by introducing queuing places, places that include a queue and a storage for processed jobs, to exploit the techniques for the evaluation and analysis of properties and performances on GSPN in combination with the benefits of QN. QPN define a formalism that can easily and naturally model both the hardware and the software components of computer-based systems. This adds a time characterization to places, that are named timed places, as the tokens that enter a place are enqueued and should be processed by the queue server before becoming available to output transitions. PNiQ integrates GSPN and monoclass QN with the purpose of overcoming the limitations of QN (e.g., to implement fork/join constructs) and uses model transformation towards GSPN (and flow equivalents, as in [1,15]) to enable model evaluation and analysis.

In a different perspective, Möbius [8], OsMoSys [16], SIMTHE*Sys* [11] (discussed in Sect. 3) and JMT [6] explored multiformalism modeling approaches that include the combination of PN and QN submodels, proposing basic interaction semantics between elements of the two formalisms and evaluation strategies for the resulting multiformalism models that are founded onto more general approaches.

Finally, it is worth noting that PN models have been used in literature to demonstrate properties, to allow evaluation by means of model transformations, to exploit the availability of existing tools on QN and queues: an example is provided by [2], in which GSPN are used to represent and analyze multiclass M/M/k queues; another example is provided by [10], in which GSPN are used to represent and analyze QN with blocking; in [5] Stochastic PN (SPN) are used to model multiserver QN. In this paper we use analogously PN to represent queues to discuss the semantics of QN-PN multiformalism interactions, for sake of clarity.

3   Multiformalism Approaches

Multiformalism approaches [9] (a case of multiparadigm approaches) advocate the use of multiple formalisms in modeling to exploit the most suitable formalism for each part of a model, to allow domain experts using the most natural formalisms for each context and perspective in a design, to exploit the most expressive constructs and keep models manageable by means of hierarchical and modular approaches. Model evaluation techniques may be guided by the structure of a model by exploiting heterogeneous submodels composition, as submodels are characterized by a given formalism (or formalisms combination) and define boundaries within the overall model that might be used for partial, parametric or iterative evaluation. The benefits of a multiformalism approach include composeability, that (partially) enables reuse of modeling artifacts, separation of concerns, collaboration between domain experts of different subsystems represented by submodels focusing on different aspects of the systems to be analyzed, readability and mantainability of complex models.

In this paper we consider QN-GSPN multiformalism models. As GSPN are Turing-equivalent, one may object that GSPN are sufficient to represent all cases: such an approach would not guarantee the benefits of multiformalism modeling, as GSPN do not natively support modularity, nor decomposition-based evaluation techniques, and suffer of the state space explosion problem.

The evaluation of multiformalism models is a challenge. In the following, we will refer to the process of evaluating a model or a submodel with *solution process*, and consequently we will define *solving* the action of evaluating a model and *solver* a tool for the evaluation of a model.

When multiformalism is not managed (that is, substantially removed) by means of model transformations, thus generating a new comprehensive model from two submodels speficied by means of different formalisms, such submodels keep their specific nature and must interface with proper mechanisms: for example, they can exchange parameters, as a result of partial solution of submodels, as in OsMoSys [16], and such parameters are used by submodels to get to a state in which they can be solved, by applying a proper solution strategy (a workflow, as in OsMoSys, or an algorithmic description of the solution process); or they can interface with each other by means of *state superposition* and/or *event synchronization* mechanisms as in Möbius [8] or SIMTHE*Sys* [11]. A state superposition mechanism operates by synchronizing two state conditions, one from each submodel, in order to obtain that each change of those states (that are actually a single state of the overall model) reflects simultaneously on both the submodels and is influenced by the conditions of both the submodels (to give an analogy, one might think of such a mechanism as a variable that is shared between two concurrent processes); an event synchronization mechanism operates by simultaneously changing the state of both the submodels when any of the synchronized events occur because of the proper mechanism of the submodel that triggers the event, obtaining a concurrent evolution of the two submodels each in its own state space (to give an analogy, one might think of such a mechanism as a barrier construct in concurrent programming). These *state* and *event* concepts are here considered as abstractions that generalize any construct that has the role of describing a stable condition, completely or as an independent component of its description (*state variable*), and has the role of allowing a change of condition, respectively, in the various possible formalisms that may be used in a multiformalism model.

3.1   An Example: QN-PN Multiformalism with Superpositions

Let us consider a QN queue in these terms, for example: the current number of jobs in the queue is described by a state variable, the current state of the server is described by a state variable, the overall number of jobs in the queue is a state variable as well, obtained by adding the two previously mentioned variables, and the end of the service is an event. In case of a PN, the situation is simpler: the number of tokens in a place is a state variable, the firing of a transition is an event.

As a consequence, in this framework, natural interactions between QN and PN see the possibility of having a queue server conceptually superposed to a transition or a place can be conceptually superposed to a queuing station. The event of the end of processing of a job in the server might thus produce tokens in a place, and the firing of a transition can generate jobs in the queuing station. In the left part of Fig. 1, the upper part represents a situation in which the completion of the service of a job generates a token in a place, with the same semantics of a firing event of a transition, conceptually depicted in the lower part; analogously, at the right, the firing of a transition generates a new job in a queue, with the same semantics of an arrival, as the queue were semantically equivalent to a place.

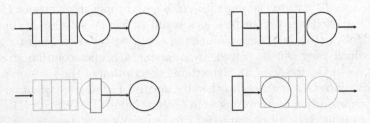

Fig. 1. Conceptual representation of the natural composition of PN and QN.

For a complete description of a multiformalism approach based on Finite Capacity Queuing Networks (FCQN) and SPN, and an example of application, the reader can refer to [12].

3.2   Some Non-trivial Cases

While the state and event superposition mechanisms is a powerful tool to define and understand multiformalism semantics, there are other interesting cases that go beyond this approach and can exploit a different paradigm, that is the foundation of the SIMTHE*Sys* multiformalism framework. In SIMTHE*Sys* the interactions between formalisms are based on the definition of formalism elements that can assume the behavior of elements belonging to different formalisms, exploiting the SIMTHE*Sys* metamodeling hierarchy to enable interactions based on the native semantics of each formalism [11]. In fact, there is a number of typical cases that arise when modeling real situations that would benefit of a multiformalism representation but are not easily captured by the situations shown in Fig. 1, unless choosing a convoluted PN representation that would let the immediateness of multiformalism representations lost. For example, well known situations, such as the characterization of an impatient customer in a queue that decides to quit after a while, the eviction of a job from a queue, a failure in processing a job, a time out during queue operations may be conveniently (and immediately) representable as a timed transition consuming a job from the queuing section or the service section of a queue (as in the left part of Fig. 2); similarly, the need for using a resource of the server section of a queue in order to process a job may be represented as a place with an incoming arc directed to the queue, so that the server section of the queue would consume a token per service (as in the right part of Fig. 2); analogously, mechanisms that turn on and off a server section in a queue, or enable and disable it, or that can temporarily reduce the number of servers or server capacity, may be immediately understood by domain experts if represented like in Fig. 3, in which the presence (in the left part) or the absence (in the right part) of one or more tokens in the place may enable the server section of the queue or influence its behavior by means of a test arc or a inhibitor arc.

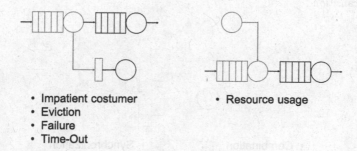

• Impatient costumer
• Eviction
• Failure
• Time-Out

• Resource usage

Fig. 2. Advanced connection types: queue to transition and place to queue.

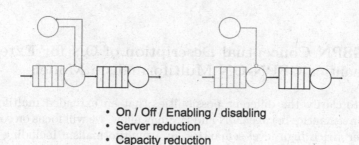

• On / Off / Enabling / disabling
• Server reduction
• Capacity reduction

Fig. 3. Advanced connection types: test and inhibitor arcs.

Other interesting cases include mechanisms that are often used in real situations that may be not completely modeled by means of QN: for example, overflow buffers that store items in assembly lines when there is no room in the queue dedicated to a machine (as in the left part of Fig. 4), that can be modeled with a place acting as buffer before a queue; relay mechanisms between two machines (as in the right part of Fig. 4), that can be modeled by means of a timed transition between two queues; situations in which there is an assembly of different parts, that produce a new item as a combination of previously independent ones (as in the left part of Fig. 5, in which a token is produced by a timed transition consuming two processed jobs from two queues); a synchronization of machines (as in the right part of Fig. 5, in which a token is produced by an immediate transition as soon as two processed jobs from two queues are available).

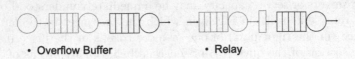

• Overflow Buffer   • Relay

Fig. 4. Advanced connection types: exclusive connection from place to queue and from queue to transition.

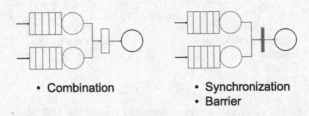

• Combination   • Synchronization
                         • Barrier

Fig. 5. Advanced connection types: from several queues directly to a timed or immediate transition.

4   A GSPN Conceptual Description of QN for Extended Semantics GSPN-QN Multiformalism Models

In order to clarify the different possibilities that an extended multiformalism interaction semantics between QN and PN may allow, we will focus on some cases which offer a rich high level semantics, in a multiformalism including QN and GSPN. We introduce GSPN equivalents of QN ../G/C/K nodes to show the exact meaning of the interactions on the conceptual level. As our goal is not to define the most compact equivalent GSPN but structures that are suitable to exactly represent the entry points that match the actual correspondence between the QN and GSPN nodes as elements of our multiformalism modeling framework, the proposed GSPN should not be considered as actual implementations, and their representations include, in a different color, arcs that represent the connections to elements of other GSPN submodels representing QN nodes in a larger model.

Figure 6 presents a GSPN equivalent of a QN node (queue i in an hypothetical QN model) with non-preemptive scheduling policy, finite capacity K and C servers. Place $OnEnter_i$ represents the situation in which a job is sent to queue i from other input queues (in the figure, hypothetical queues i-$1A$ and i-$1B$). A token produced in this place may alternatively enter the queue if there is at least one free position (by a firing event of transition $Enter\_Station_i$) or be lost (by a firing of transition $Loss_i$): these transition are respectively enabled if there is at least one token or if there is no token in place $Capacity_i$, that is initialized with K tokens to describe a queue with finite capacity K.

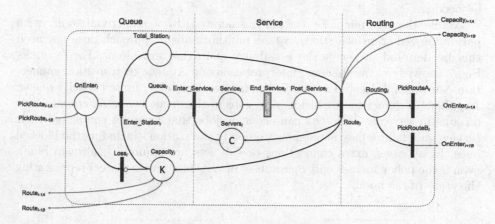

Fig. 6. Conceptual representation of a QN node with non-preemptive scheduling policy using GSPN.

When $Enter\_Station_i$ fires, a token is added in place $Queue_i$, that represents the jobs currently in the queue, and in place $Total\_Station_i$, that counts the jobs currently in the queue. In figure, these elements form section $Queue$. Section $Service$ represents the processing of jobs by the queue servers.

Place $Servers_i$ represents the currently available servers (initialized to C) in the queue: when $Enter\_Service_i$ fires, tokens from $Queue_i$ and $Servers_i$ are consumed and an equal number of tokens is produced in $Service_i$ to represent the jobs currently processed in the busy servers. The C-server timed transition $End\_Service_i$ represents the processing operation, with exponentially distributed service times; place $Post\_Service_i$ stores processed jobs, so that transition $Route_i$ can update the total number of jobs in the queue (marking of $Total\_Station_i$), the total number of currently free servers (marking of $Servers_i$) and the current available capacity of the queue (marking of $Capacity_i$) when entering section $Routing$ of the figure. The same transition produces a token in place $Routing_i$ per processed job, so that one of the $PickRouteX_i$ transitions, one per destination output queue in a complete model (in figure, $PickRouteA_i$ and $PickRouteB_i$ that route jobs toward two hypothetical queues $i+1A$ and $i+1B$), can consume the token representing the job sent to the related output queue, producing it in its equivalent of $OnEnter_i$.

Place $Capacity_i$ may also be used to influence the enabling of the equivalents of transition $Route_i$ of input queues (in the figure, hypothetical queues $i-1A$ and $i-1B$) to describe a FCQN without loss.

While one may be tempted to consider a parametric version of this net (or an equivalent one) to directly evaluate models by model transformation, an approach which is general enough to allow the implementation of comprehensive and highly parameterizable and automated tools (such as JMT), aiming at supporting domain experts and modelers with a rich set of options, is needed. In fact, with a simple change of policy of the server, a different structure of the net may be needed.

While this net might be certainly generated, by a non-trivial tool[2], with parameterized transformations, a true multiformalism approach does not need this burden and preserves the possibility of further extensions. For example, Fig. 7 shows how the different interpretations of a queue to transition connection can be implemented for a non-preemptive scheduling finite capacity queue: *Impatient costumer* must consider only jobs in the queue, while *Eviction* focuses on jobs in services, *Time-Out* can occur on jobs that have not completed their service, and failure may happen on any job in the station - including the blocked ones. In all cases, extra connections or sub-nets are required (shown in blue), even if the policy focuses only on a subset of variables (green places) representing the state of the node.

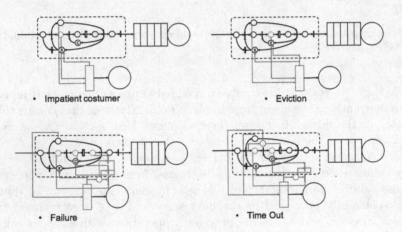

Fig. 7. Implementation of the different types of Queue to Transition connections. (Color figure online)

5   A Running Example

To introduce the need for new interconnection semantics, we will use a running example. It consists in a system, represented in Fig. 8, that is packaging and delivering goods according to incoming orders. The system consists of stages that operate as an assembly line: there is a *Process and Packing* stage, which

---

[2] For example, similarly as in [17].

collects orders and consequently selects and enqueues goods to be dispatched, acquires packages, packs each piece of goods and is resupplied with batches of v_1 packages each time a batch terminates, and a *Shipping* stage, in which packaged pieces are enqueued to be loaded on a *Truck* and be delivered. If items wait too much for shipping, a proper mechanism moves them in another section of the *Shipping* stage that is an alternative, auxiliary shipping line, which is prone to failures. A *Truck* is full and leaves when v_2 items are loaded.

We are interested in modeling performances of the system. As in stages *Process* and *Shipping* the prevalent aspect is the enqueuing process, while in the *Packaging* stage the prevalent aspect is the acquisition of the packaging and the related packaging operation, this system is suitable for a multiformalism model, that is represented in Fig. 9, in which only the most significant elements are labeled for the sake of brevity. Anyway, the extraction of items waiting for too much time in the main shipping facility is not supported by ordinary semantics of queuing networks, so it cannot be easily represented with the natural QN-PN multiformalism semantics. A first approximated model exploiting the semantics presented in Fig. 1 can be defined, neglecting the additional shipping line, to have a rough evaluation of performances.

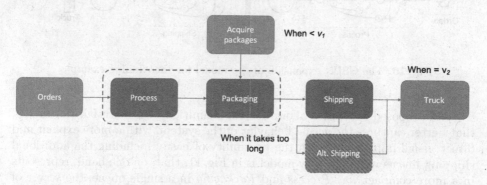

Fig. 8. A running example: goods package and delivery.

The *Process* stage is thus represented by a queue that receives orders by a source, while the *Packaging* stage is represented by a PN, the *Shipping* stage is

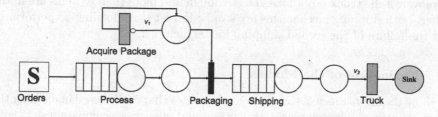

Fig. 9. The approximated multiformalism model for the running example.

represented by a queue and the *Truck* stage is represented by a PN (all PN transitions are timed transitions with exponentially distributed service times). The *Packaging* PN consists of a section that describes the acquisition of a package (in the top part of the figure) and a part that models the packaging operation when a piece of good and its packaging are ready to be assembled (that is, the tokens that represent them are present in the input places of immediate transition *Packaging*). We can conceptually expand this model by ideally replacing queues with the first equivalent net presented in the previous Section, obtaining a GSPN model like in Fig. 10, considering that both the queues, that represent real components of a physical system, have finite capacity and should operate with a non-preemptive scheduling policy (once one packaging operation starts, it should be completed, and one package per time can be prepared).

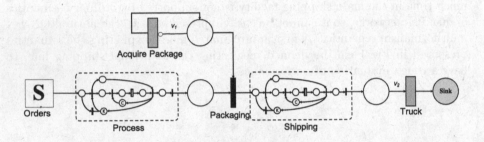

Fig. 10. The GSPN expansion of the model for the running example.

The proposed semantics allows for the definition of a more faithful model, that better captures the actual behavior of the system, with a more explicit and direct visual impact and with the possibility of easily including the additional shipping line as well. The new model is in Fig. 11, that, on one hand, represents in a more compact way *Process* and *Packaging* in a single queue, the server of which can process a job only if a token is available for consumption from the PN in the upper part of the figure, representing the availability of the packaging to be used, and, on the other hand, allows a more complex representation of *Shipping*, that now includes a second queue (*Alt.Shipping*) with *Shipping* and a relay mechanism (*LongDelay*) that moves an item from the first queue to the new queue to relief its workload when operations are too slow. In this version, moreover, if the truck is not loaded fast enough and more than v_2 items are in the buffer, both shipping mechanisms are stopped to let truck loading be performed, and the failure of the second shipping line is included.

5.1   Evaluation of the Models

To show the importance of the different semantics that can be used in defining the interconnections between places, transitions and queues, we compare results with 5 different versions of the system modeled in Fig. 8, namely: A) the simplified

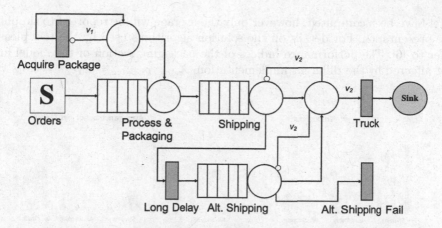

Fig. 11. The improved multiformalism model of the system, with packaging integrated with processing, alternative shipping, blocking and failure.

model in Fig. 9, B) the complete model in Fig. 11, C) case B without failure (i.e. missing transition *Alt. Shipping Fail*), D) case B without blocking (that is, without the inhibitor arcs starting ending on *Shipping* and Alt. Shipping), E) case B where also jobs in service at the shipping might be shifted to the alternate service if the timeout expires (i.e. using the *Time Out* block of Fig. 7 instead of the *Impatient costumer* one). This means that jobs in service can be stopped and restarted on the alternate server. All models share the same timing parameters shown in Table 1, and all events are considered to occur after an exponentially distributed amount of time.

Table 1. Model parameters

Event	Average duration
Acquire package	0.8 min
Process & packaging	1 min
Shipping	1.2 min
Truck	7.5 min
Long delay	10 min
Alt. Shipping	2.5 min
Alt. Shipping fail	50 min
v_1	5
v_2	10

Results have been computed in JMT, using PN expansion to consider connections not supported yet by the tool. 99% confidence intervals, with 3% confidence level have been computed, however only the average will be reported to simplify the presentation. For details on the solution algorithms used by the tool please refer to [6]. The performance indices of the packaging section of the model are not affected by the different implementation, $X_{Acq.\ Package} = X_{P\&P} = 0.5\ s^{-1}$..

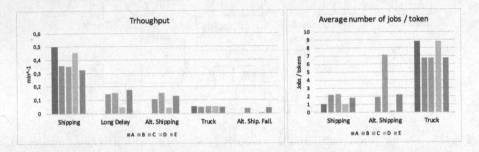

Fig. 12. Different results for the same system, under different assumptions and semantics.

Figure 12 shows the results for the case where it is more visible the impact of the different semantic being used. Case A, although being able to correctly model the packaging part, lacks of details, and produces a large deviation especially in the average occupations. When failure is not considered (case C), the system enjoys the highest throughput, at the expense of larger queues which result in a higher latency. If no blocking is considered (case D), the alternate shipping service is less used, resulting in longer queues, partially compensated by a reduction in failure states. The effect of using Time-Out instead of Failure semantic (case E) increases the load of the alternate shipping circuit, leading to longer response times and global failure probability due to the increased possibility of being transferred to the secondary shipping method. Although scenarios A, C, D and E could be valid system alternatives, if the one under study corresponds to case B, the use of proper semantic can lead to different results, that might lead to wrong resource planning. For this reason, formalism clarifications like the one proposed in this work become of paramount importance for producing accurate performance estimates.

6   Conclusions and Future Work

The richness of the formalism definitions for QN and PN and the flexibility offered by multiformlism modeling provide much more expressive power and semantic possibilities than literature explored and may enable a higher degree of user friendliness of these popular formalisms for domain experts. In this paper we just presented a set of interesting cases which may be expressed by means of new interactions between elements of the QN and PN formalisms. We plan to further develop the investigation on the open perspectives and to study the actual implementation of the proposed solutions in a consolidated tool like JMT. We also plan to exploit the SIMTHE*Sys* multiformalism modeling framework to support automatic state space and transition matrix generation for analytical solution when events are exponentially distributed. Our aim is to provide an extended and consistent support for QN-PN multiformalism models implementation in a consolidated and well known tool to further encourage its diffusion and to promote the culture of model-based quantitative performance evaluation in industrial environments, while advancing the understanding of QN, PN and QN-PN multiformalism on the methodological level, investigating additional aspects such as colored/multiclass models, blocking policies, arc weight semantics, complex management of finite capacity features, job/token conservation, fork/join like constructs, job/token tracing, and novel performance indices exploiting the new features.

References

1. Balbo, G., Bruell, S., Ghanta, S.: Combining queuing network and generalized stochastic Petri Net models for the analysis of a software blocking phenomenon. In: International Workshop on Timed Petri Nets, pp. 208–225 (1985)
2. Balsamo, S., Marin, A.: On representing multiclass M/M/K queues by generalized stochastic Petri Nets. In: Proceedings of ASMTA 2007: 14th International Conference on Analytical and Stochastic Modelling Techniques and Applications, pp. 121–128 (2007)
3. Bause, F.: Queueing Petri Nets - a formalism for the combined qualitative and quantitative analysis of systems. In: 1993 Proceedings of 5th International Workshop on Petri Nets and Performance Models, PNPM, pp. 14–23 (1993)
4. Becker, M., Szczerbicka, H.: PNiQ: integration of queuing networks in generalised stochastic Petri Nets. IEE Proc. Softw. **146**(1), 27–32 (1999). https://doi.org/10.1049/ip-sen:19990153
5. Boon-in, P., Vatanawood, W.: Formal modeling of multi-server queuing network using Stochastic Petri Nets. In: ACM International Conference Proceeding Series, pp. 44–50 (2019). https://doi.org/10.1145/3374549.3374563
6. Casale, G., Serazzi, G., Zhu, L.: Performance evaluation with java modelling tools: a hands-on introduction. Perform. Eval. Rev. **45**(3), 246–247 (2018). https://doi.org/10.1145/3199524.3199567
7. Lara, J., Vangheluwe, H.: AToM^3: a tool for multi-formalism and meta-modelling. In: Kutsche, R.-D., Weber, H. (eds.) FASE 2002. LNCS, vol. 2306, pp. 174–188. Springer, Heidelberg (2002). https://doi.org/10.1007/3-540-45923-5_12
8. Deavours, D., et al.: The Möbius framework and its implementation. IEEE Trans. Softw. Eng. **28**(10), 956–969 (2002). https://doi.org/10.1109/TSE.2002.1041052
9. Gribaudo, M., Iacono, M.: An introduction to multiformalism modeling. In: Theory and Application of Multi-formalism Modeling, pp. 1–16 (2013). https://doi.org/10.4018/978-1-4666-4659-9.ch001
10. Gribaudo, M., Sereno, M.: GSPN semantics for queueing networks with blocking. In: International Workshop on Petri Nets and Performance Models, pp. 26–35 (1997)
11. Iacono, M., Barbierato, E., Gribaudo, M.: The SIMTHESys multiformalism modeling framework. Comput. Math. Appl. **64**(12), 3828–3839 (2012). https://doi.org/10.1016/j.camwa.2012.03.009

12. Iacono, M., Gribaudo, M.: Element based semantics in multi formalism performance models. In: MASCOTS, pp. 413–416. IEEE (2010)
13. Kounev, S., Dutz, C.: QPME: a performance modeling tool based on queueing Petri Nets. SIGMETRICS Perform. Eval. Rev. **36**(4), 46–51 (2009). https://doi.org/10.1145/1530873.1530883
14. Lindemann, C., Hommel, G.: Combining deterministic and stochastic Petri net and product-form queueing network models for evaluating gracefully degradable systems. In: Proceedings, Advanced Computer Technology, Reliable Systems and Applications, pp. 880–884 (1991)
15. Menascé, D.: A methodology for combining GSPNs and QNs. In: 37th International Conference Computer Measurement Group (2011)
16. Moscato, F., Flammini, F., Di Lorenzo, G., Vittorini, V., Marrone, S., Iacono, M.: The software architecture of the OsMoSys multisolution framework. In: VALUETOOLS 2007–2nd International ICST Conference on Performance Evaluation Methodologies and Tools (2007)
17. Raiteri, D., Iacono, M., Franceschinis, G., Vittorini, V.: Repairable fault tree for the automatic evaluation of repair policies. Proceedings of the International Conference on Dependable Systems and Networks, pp. 659–668 (2004). https://doi.org/10.1109/dsn.2004.1311936

Author Index

Printed in the United States
by Baker & Taylor Publisher Services